Lecture Notes in Computer Sc

Edited by G. Goos and J. Hartmanis

Advisory Board: W. Brauer   D. Gries   J. Stoer

Mark de Berg

Ray Shooting, Depth Orders and Hidden Surface Removal

Springer-Verlag

Berlin Heidelberg NewYork
London Paris Tokyo
Hong Kong Barcelona
Budapest

Series Editors

Gerhard Goos
Universität Karlsruhe
Postfach 69 80
Vincenz-Priessnitz-Straße 1
D-76131 Karlsruhe, FRG

Juris Hartmanis
Cornell University
Department of Computer Science
4130 Upson Hall
Ithaca, NY 14853, USA

Author

Mark de Berg
Department of Computer Science, Utrecht University
Padualaan 14, NL-3508 TB Utrecht, The Netherlands

CR Subject Classification (1991): I.3.5-8, I.4.8

ISBN 3-540-57020-9 Springer-Verlag Berlin Heidelberg New York
ISBN 0-387-57020-9 Springer-Verlag New York Berlin Heidelberg

This work is subject to copyright. All rights are reserved, whether the whole or part
of the material is concerned, specifically the rights of translation, reprinting, re-use
of illustrations, recitation, broadcasting, reproduction on microfilms or in any other
way, and storage in data banks. Duplication of this publication or parts thereof is
permitted only under the provisions of the German Copyright Law of September 9,
1965, in its current version, and permission for use must always be obtained from
Springer-Verlag. Violations are liable for prosecution under the German Copyright
Law.

© Springer-Verlag Berlin Heidelberg 1993
Printed in Germany

Typesetting: Camera ready by author
45/3140-543210 - Printed on acid-free paper

Preface

Computational geometry is the part of theoretical computer science that concerns itself with geometric objects; it tries to design efficient algorithms for problems involving points, lines, polygons, and so on. The field has gained popularity very rapidly during the last decade. This is partly due to the many application areas of computational geometry and partly due to the beauty of the field itself. I hope that the reader will experience some of this beauty when reading this book. Initially, most research was directed towards problems in two-dimensional space, and good solutions to many of these problems have been found. In recent years the field has shifted its attention to problems in three- (and higher) dimensional space. This shift of attention is also reflected in this book: almost everything that is studied is three-dimensional.

This book focuses on three problems in computational geometry that arise in computer graphics. (A more ample discussion of the relation of these problems to computer graphics can be found in Chapter 1.) The first problem is the ray shooting problem: preprocess a set of polyhedra into a data structure such that the first polyhedron that is hit by a query ray can be determined quickly. The second problem is that of computing depth orders: we want to sort a set of polyhedra such that if one polyhedron is (partially) obscured by another polyhedron then it come first in the order. The third problem that we study is the hidden surface removal problem: given a set of polyhedra and a view point, compute which parts of the polyhedra are visible from the view point. These are not only three nice problems that arise naturally in one of the application areas of computational geometry; they involve issues that are fundamental to three-dimensional computational geometry and are, hence, also of considerable theoretical interest. The book also contains a large introductory part, which discusses the techniques that will be used to tackle the three problems stated above. This part should be interesting not only to those who want to read the rest of the book but miss the necessary background, but to anyone who wants to know more about some (recent) techniques in computational geometry.

This book is a revised version of my Ph.D. thesis, which was the result of the research I did at the department of computer science of Utrecht University (the Netherlands) in the period from August 1988 to March 1992. The research was supported by the Dutch Organization for Scientific Research (N.W.O.), and partially by the ESPRIT Basic Research Action No. 3075 (project ALCOM). There are many people without whom this book would not have been what it is now. First of all, there is Mark Overmars, who introduced me to the fascinating field of computational geometry: I could not have wished myself a more stimulating supervisor for my Ph.D. research. Secondly, there is Marc van Kreveld, my roommate during all those years:

they would certainly have been less pleasant without the many discussions we had on various subjects (including computational geometry). Practically everything in this book has profited from their comments and suggestions. Let me also mention the other people that contributed to the research that is reported in this book: Hazel Everett, Danny Halperin, Otfried Schwarzkopf, Jack Snoeyink and Hubert Wagener. Finally, I thank Jan van Leeuwen, Kees van Overveld, Micha Sharir and Emo Welzl for being in the reading committee of my thesis.

Utrecht, April 1993   Mark de Berg

Contents

Part A

Introduction

Teil 1

Einführung

Chapter 1

Computational Geometry and Computer Graphics

This book studies three problems in computational geometry that arise in computer graphics. Computational geometry is the part of theoretical computer science that concerns itself with geometric objects; it tries to design efficient algorithms for problems involving points, lines, polyhedra, et cetera, in one-, two- and more-dimensional space. A paper [114] by Michael Shamos in 1975 marks the beginning of computational geometry as a separate area in algorithms research. Since then computational geometry has attracted many researchers and it has made tremendous progress. By now it has established itself as an important field within theoretical computer science: hundreds of people are working in the field, thousands of papers have been published, every major conference or journal on theoretical computer science has papers on it, and there are even several journals and conferences which are devoted solely to computational geometry.

What is it that makes computational geometry so appealing? The reason for this is twofold. First of all, most problems have a simple and intuitive definition and are easily visualized, which makes it easy to interest people in them. In the second place, the applications of computational geometry are numerous, in particular in areas like computer graphics, robotics, geographical information systems (GIS), databases, and VLSI design. Computational geometry offers the right abstraction to study the problems that arise in these areas. Let us give a few prime examples of these applications.

Consider a database that stores information about, say, the age and number of children of people. A typical query in this database is of the form: "Which persons are between 20 and 65 years of age, and have between 13 and 18 children?" If we represent each person by a point in a two-dimensional space, where the first coordinate of the point is the age of the person and the second coordinate is the number of children, then the query asks for all points in the rectangle $[20 : 65] \times [13 : 18]$. See Figure 1.1. This problem is called the *orthogonal range searching problem* in computational geometry, and it has been studied extensively.

A second example can be found in what are called *motion planning problems* in computational geometry. Who did not have the frustrating experience of a sofa that

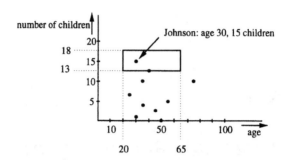

Figure 1.1: A query in a database.

refuses to leave through the same door that it was brought in through? Planning the motion of an object from a starting position to a goal position is an essential problem in robotics. Motion planning problems have been studied in great detail in computational geometry, leading to papers with intriguing titles like "How to Move a Chair through a Door"[126] and "On the Piano Movers' Problem, I: The Case of a Two-Dimensional Rigid Polygonal Body Moving amidst Polygonal Barriers"[113].

A third important application area is computer graphics. Since the problems that are studied in this book originate in this area, we discuss it in more detail.

Computer graphics concerns itself with the display of visual information on the screen of a computer terminal. We can roughly subdivide computer graphics into two subareas. One considers hardware aspects, and the other studies algorithmic aspects. It is the latter area where computational geometry comes into play and which has our interest. There is an enormous amount of literature on computer graphics, and we will not even try to give a survey. We just mention two good textbooks, by Foley et al. [58] and by Watt [120], which the interested reader can consult.

The most fundamental algorithmic problems arise when one wants to display the view of a three-dimensional scene. As an example, consider an architect who is designing a building. It would be useful for her to see the building before it is actually constructed. This can be accomplished using computer graphics. The architect tells the system exactly where walls, doors, windows, tables, chairs and other objects are located, what their shape is, and so on, and the system calculates what the building looks like for an observer standing, say, at the main entrance. The system might even compute what the building looks like from the inside, giving the architect the possibility to walk through the building, so to speak. Thus the system has to compute the view of a collection of objects in three-dimensional space as seen from a certain view point. A simple illustration is given in Figure 1.2. The problem of determining which parts of each object are visible and which parts are hidden is called the *hidden surface removal problem*.

To understand the different approaches to the hidden surface removal problem, one has to know how a single, completely visible, object is displayed onto the screen. A computer screen consists of many small dots (typically about one million) called

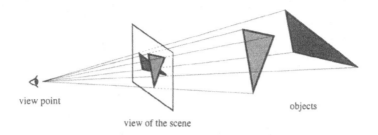

Figure 1.2: Viewing two triangles in three-dimensional space.

pixels. To display an object it is projected onto the screen with the view point as the center of projection. The projected image of the object covers a number of pixels on the screen. These pixels get the color of the object, and the remaining pixels get the background color. A realistic scene consists of many objects, and a pixel can be covered by several of them. It is for the computer to decide which object is visible at each pixel, so that it can give the pixels the right color. Hidden surface removal algorithms are used to compute which object is visible at which place. They are usually classified into *image space algorithms*, which work with the projected images of the objects, and *object space algorithms*, which work with the objects themselves.

Image space algorithms typically look for every pixel on the screen at all the objects whose projected image covers the pixel; the one that is closest to the view point is displayed. A notable example of such a method is the *depth-buffer algorithm*, also called *z-buffer algorithm*. This algorithm works as follows. It processes each object in turn, maintaining a buffer that stores for every pixel the (depth of the) currently visible object. To process a new object, one tests for every pixel that lies in the projected image of the object whether the new object is closer to the view point (at that pixel) than the currently visible object. If this is the case, then the buffer is updated. The main advantage of this method is that it is easily implemented in hardware. Therefore the method is—despite its brute-force approach—fast in practice and often used.

An interesting variation on this method is the *depth sorting algorithm*. This method eliminates the test between the processed object and the currently visible object in the depth-buffer algorithm. Moreover, the algorithm does not need a buffer for maintaining the z-values for each pixel. It works in two phases. In the first phase of the algorithm the objects are sorted according to their distance from the view point: if an object A is (partially) obscured by object B then A comes before B in the ordering. Such an order on the objects is called a depth order. Note that a depth order on the objects does not always exist, since there can be *cyclic overlap* among the objects. See, for example, Figure 1.3. In such cases the cycles have to be broken by cutting the objects into smaller pieces. This first phase works in object space. Efficient algorithms for computing depth orders are presented in Part C of this book. The second phase is similar to the depth buffer method, i.e. the objects are drawn

Figure 1.3: Three triangles with cyclic overlap.

one by one onto the screen. Because the order in which the objects are processed is a depth order—that is, objects in the back are processed earlier than objects in the front—none of the objects already processed can obscure a new object. Hence, all pixels in the projected image should get the color of the new object and no depth comparison is needed. Note that the second phase of the depth order algorithm is similar to the way an artist makes a painting: first the background colors are painted and later the objects are painted 'on top of this'. Thus the algorithm is sometimes called the *painter's algorithm*.

Another way of computing which object is visible at a certain pixel is to 'shoot a ray from the view point through the pixel'. The first object that is hit is the one that is visible at that pixel. See Figure 1.4. Algorithms following this approach are

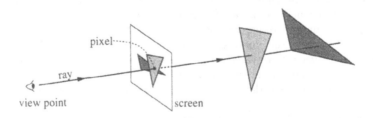

Figure 1.4: Hidden surface removal using ray tracing.

called *ray tracing algorithms*. Glassner [61] discusses this technique in great detail. To speed up the tracing process, the objects are stored in a data structure that allows one to determine the first object that is hit without looking at all the objects. The problem of designing efficient data structures for this task is called the *ray shooting problem* in computational geometry. It is the topic of Part B of this book.

Obtaining realistic pictures of a scene not only involves computing which parts of the objects can be seen from the view point, but also computing shading information about the light intensity that reflects from an object. The ray tracing algorithm can be extended to provide this information. The idea of the method is as follows. By shooting a so-called *primary ray* from the view point through each pixel, one can determine which object, let's call it A, is visible at that pixel. Let p be the point where the primary ray through the pixel hits A. To compute the shading information we shoot extra rays, called a *shadow feelers*, from point p into the direction of each light source in the scene. If a shadow feeler does not hit any object before it reaches the light source, then we know that p is lit by the light source. Otherwise p will be in shadow. It is, however, still possible that p is lit indirectly by the light source, by reflections via other objects. Such indirect illumination effects can also be computed, by shooting even more rays. It is beyond the scope of this book to discuss the ray tracing method any further, but it should be clear that the method requires many rays to be traced. Hence, it is extremely important to be able to compute the first object hit by a ray efficiently. The design of data structures for this task is the topic of Part B of this book.

The pixels on the screen can be grouped into regions where the same object is visible. The image space methods discussed above compute the view of a scene pixel by pixel. This means that the 'structure' of the view is lost. Object space algorithms compute a combinatorial representation of the structure of the view. Let us be more precise about this. The view of a scene is a subdivision of the viewing plane into maximal connected regions in each of which (some portion of) a single object can be seen, or no object is seen. In Figure 1.4, for example, this subdivision consists of four regions; the light triangle is visible in one of them, the dark triangle is visible in two other regions, and no object is seen in the fourth, white region. Object space algorithms compute this subdivision—which is called the *visibility map* of the given set of objects—as a collection of (polygonal) faces. In other words, object space algorithms compute exactly which parts of each object are visible. After that the visible parts can be projected and displayed without difficulty. Part D is devoted to the computation of these maps.

In general, object space hidden surface removal methods tend to be slower than image space methods such as the z-buffer algorithm, because they cannot be implemented in hardware very well. However, object space algorithms have certain advantages over image space methods. Suppose that we want to display the hidden lines in a scene dashed, instead of making them invisible. Object space algorithms compute exactly which parts of each line are visible and which parts are not, thus making it easy to display the invisible part dashed. Image space algorithms do not provide the information that is necessary to achieve this. Another weak point of image space algorithms comes up when one wants to print the view of a scene on paper, instead of displaying it on the screen of a computer terminal. When hidden surface removal has been done in image space, the only thing we can do is to plot every pixel separately. But this method fails to take advantage of the fact that the resolution (that is, the number of pixels per square inch) of modern laser printers is much higher than the resolution of computer screens. If hidden surface removal has been performed in object space then the visibility map can be processed directly, resulting in a picture of higher quality. A third advantage of object space algorithms is that they can be used to compute shadows in a scene. This follows from the fact that a

point on an object is lit by a light source if and only if the ray from the light source to
that point does not intersect any other object on its way. In other words, the part of
a scene that is lit by a light source is exactly the same as what can be seen from the
light source. Thus we can use an object space hidden surface removal algorithm to
compute which parts of the objects are lit by the light source. Image space solutions
such as the z-buffer method do not have this possibility: we can apply a z-buffer-like
algorithm to compute the 'view' from the light source, but the problem is that the
pixels with respect to the viewpoint do not coincide with the 'pixels' with respect to
the light source. Notice that the shadows in a scene do not change when the view
point moves. (Although what is visible of a shadow can change, of course.) Hence, the
combinatorial representation of the shadow—which is exactly what an object-space
hidden surface removal algorithm can compute—can be used for every view point.
If shadow calculation is performed using the ray tracing algorithm described above,
then everything has to be computed anew for each view point.

Before we give an overview of the contents of this book, let us spend a few words
on the status of our work. This is a book about problems in computational geometry
that arise in computer graphics, not a book about computer graphics. Hence, we give
a theoretical treatment of these problems: we derive exact time bounds for our algorithms that will show that our algorithms perform better (in theory) than previously
known algorithms. On the other hand, we have not implemented any of our methods.
Thus the applicability in practice has not been established yet. Indeed, some of our
data structures probably will not perform well in practice if they are implemented
exactly as described here. Clearly, this is a topic of future study. In any case, a
theoretical study such as undertaken here provides us with a deeper understanding
of the important issues involved in the problems, which can be useful to obtain good
practical solutions.

The book consists of four parts.

Part A, of which this introduction is the first chapter, introduces the reader to
the problems that are studied and to the techniques that we use to solve them. Actually, this part might be interesting not only to people who want to read this book,
but to anyone who wants to know more about some of the (recent) techniques in
computational geometry.

Part B concerns itself with the ray shooting problem: Preprocess a set of polyhedra
in three-dimensional space into a data structure, such that the first polyhedron that
is hit by a query ray can be computed efficiently. We develop new, efficient data
structures for various settings of this problem. In particular, we distinguish the case
where the origin of the rays is fixed, the case where the direction of the rays is fixed,
and the general case where there are no restrictions on the ray. For each setting,
we study a number of different scenes: scenes consisting of axis-parallel polyhedra
(whose edges are parallel to the x-axis, the y-axis, or the z-axis), scenes consisting
of c-oriented polyhedra (whose edges are parallel to c different axes), and scenes
consisting of arbitrary polyhedra.

In Part C we study the problem of computing depth orders: Given a set of
polygons, sort them according to their distance to the view point. We start by
studying the problem in two-dimensional space. This is useful, because many three-

dimensional scenes—in particular certain types of landscapes, which are called *polyhedral terrains*—can be treated as being two-dimensional. We present a data structure such that a depth order for a given view point can be calculated efficiently. We also study three-dimensional scenes; here we present the first algorithm to compute a depth order for a set of polygons in three-dimensional space that achieves a subquadratic running time.

Finally, in Part D we present an object space algorithm for hidden surface removal in general polyhedral scenes. The running time of our algorithm is *output-sensitive*, that is, the running time decreases as the complexity of the visibility map decreases. Previously, such algorithms were known only for special cases, where the objects in the scene satisfy the—not very realistic—constraint that a depth order on the objects exists and is known.

Chapter 2

Preliminaries

This chapter gives the basic definitions and offers a brief description of a number of techniques that have been developed in computational geometry and that are used throughout this book. The aim is not to give a complete survey of these techniques in their most general and sophisticated form; it is rather meant as a starting point for our later work. If the reader wants to know more, he/she should follow the pointers to the literature that are given. Before we continue, the reader's attention is drawn to the list of notations that is given on page 195.

2.1 Terminology

In this section we briefly recapitulate the main terminology related to algorithm analysis and design that is used in this book. A more detailed discussion can be found in [37].

In algorithms research, the running time of an algorithm is usually given as a function of the input size, that is, as a function of the number of 'elementary objects' in the input. The running time of an algorithm that sorts a set of reals, for example, is given as a function of the number of reals to be sorted. Thus we may say that a sorting algorithm runs in $17n \log n + 5n$ time, meaning that it performs at most $17n \log n + 5n$ elementary operations—like additions, comparisons and multiplications—to sort any set of n reals. We are mostly interested in the asymptotic behavior of the algorithms and not so much in the constant factors in the running time. To express this fact, we use the O-notation to denote running times, which is defined as follows:

> $f(n) = O(g(n))$ if and only if there are positive constants c and n_0 such that $0 \leqslant f(n) \leqslant cg(n)$ for all $n > n_0$.

The running time of the above sorting algorithm can thus be written as $O(n \log n)$. The O-notation is used to denote upper bounds on the running time of an algorithm. Lower bounds can be expressed using the Ω-notation:

> $f(n) = \Omega(g(n))$ if and only if there are positive constants c and n_0 such that $0 \leqslant cg(n) \leqslant f(n)$ for all $n > n_0$.

Finally, we define $f(n) = \Theta(g(n))$ if and only if $f(n) = O(g(n))$ and $f(n) = \Omega(g(n))$.

For bounds on the amount of storage we also use the O-notation. We say, for example, that a linear list that stores n reals uses $O(n)$ storage. We assume that an 'elementary object'—such as a real number or a pointer—can be stored in one unit of storage.

Above we said that the running time of an algorithm is $O(f(n))$ if it runs in $O(f(n))$ time for *any* input set of size n. Such a bound on the running time is called a *worst-case* bound. All bounds in this book are worst-case, unless stated otherwise.

Some algorithms have a good running time for an 'average' input. This is formalized as follows. Each input is assigned a certain probability that it occurs, thus giving a probability distribution over all inputs of size n. Given this distribution, we can compute the *expected* running time of the algorithm. Thus an algorithm with $O(f(n))$ expected running time runs in time $O(f(n))$ for the 'average' input. For some 'bad' inputs, however, it may take more time. We will not consider this type of algorithms in this book.

Instead, we sometimes use *randomized* algorithms. The behavior of such an algorithm is influenced by certain random choices it makes: the algorithm makes several (programmed) coin flips during its execution, and the flow of execution, and therefore the running time, depends on the outcome of these coin flips. (It is assumed here that the generation of 'random' bits is an elementary operation that takes constant time.) A typical step in a randomized algorithm is to pick a random element (or a random subset) of a set. The algorithm is based on the fact that, with high probability, this random element (subset) has some nice property that will make the algorithm run fast. This means that—in contrast to a deterministic algorithm—the running time can vary if we feed the algorithm the same input a number of times. We say that the algorithm has an $O(f(n))$ randomized running time, if the expectation of the running time over the random choices made by the algorithm is $O(f(n))$. Note that this is independent of the input: randomized algorithms do not make any assumption about the input distribution, so there are no 'bad' inputs for a randomized algorithm. Notice also that the algorithm always produces the right answer, irrespective of the outcome of the coin flips. (This is sometimes referred to as a Las Vegas-type randomized algorithm. Another type of randomized algorithms that occurs in the literature—but not in this book—is the Monte Carlo-type randomized algorithm. The running time of such an algorithm is worst-case, but the randomization may cause an error in the final answer, which is expected to be small.)

A final type of time bound that we will encounter in this book is the *amortized* time bound. Amortized analysis is used for dynamic data structures where the cost of an operation (an update or a query) can fluctuate, but where the average cost is good. Intuitively, an operation on a data structure that stores n objects is said to take $O(f(n))$ amortized time if, for any $i \geqslant 1$, the total time taken by the first i operations is $O(if(n))$. We refer the reader to [37] for a precise definition.

2.2 Geometric Definitions

In this section we give some basic geometric definitions and terminology. We assume that the reader is familiar with the most elementary geometric concepts, such as lines, line segments, hyperplanes, polygons and polyhedra. Definitions can be found in the textbook by Preparata and Shamos [104], or any standard book on geometry.

We use the standard notation $\mathbb{E}^d$ to denote d-dimensional Euclidean space. We will be mostly concerned with polyhedra in $\mathbb{E}^3$. We distinguish two special classes of polyhedra, namely axis-parallel and c-oriented polyhedra, which are defined as follows. A polyhedron is called *axis-parallel* if each edge of the polyhedron is parallel to the x-, the y- or the z-axis; a set of polyhedra is axis-parallel if all polyhedra in the set are axis-parallel. Figure 2.1(a) shows an axis-parallel polyhedron. A polyhedron is called

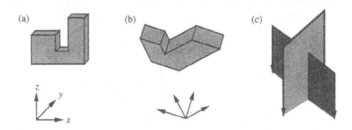

Figure 2.1: An axis-parallel polyhedron, a 4-oriented polyhedron and two intersecting curtains.

c-oriented if all of its edges are parallel to one of c fixed axes; a set of polyhedra is c-oriented if all polyhedra in the set are c-oriented with respect to the same set of c axes. (We deviate slightly from the original definition by Güting [66, 69], who defines a polyhedron to be c-oriented if its faces are parallel to c fixed planes.) For example, a set of axis-parallel polyhedra in $\mathbb{E}^3$ is 3-oriented. A 4-oriented polyhedron, together with the four axes to which its edges are parallel, is depicted in Figure 2.1(b). In the literature c is usually considered to be a constant, but there is no need to be so restrictive. Indeed, solutions for c-oriented polyhedra are often more efficient than general solutions even when c is a slowly growing function of n. For example, the structure for ray shooting from a fixed point in c-oriented polyhedra that we present in Section 5.2 is more efficient than the structure for general polyhedra presented in Section 5.3 when $c = O(n^{1/4})$.

Another class of objects that plays an important role in this book are curtains. A *curtain* is an unbounded polygon in $\mathbb{E}^3$ with three edges, two of which are parallel to the z-axis and extend downward to minus infinity. Thus the polygon can be seen as a curtain hanging from the third, bounded, edge. See Figure 2.1(c) for an example of two intersecting curtains.

Finally, we define a *polyhedral terrain* to be the graph of a piecewise linear continuous bivariate real function. Thus a polyhedral terrain consists of a connected set of polygonal faces, such that the xy-plane is the disjoint union of the orthogonal projections of these faces onto the xy-plane.

A very important concept in computational geometry are arrangements of hyperplanes. Let H be a set of hyperplanes in $\mathbb{E}^d$. The subdivision of $\mathbb{E}^d$ into connected pieces of various dimensions that is induced by this set of hyperplanes is called the *arrangement of H*, denoted by $\mathcal{A}(H)$. The pieces of dimension i are called the *i-faces* of the arrangement. The 0-faces are usually referred to as *vertices*, the 1-faces are called *edges*, the $(d-1)$-faces are called facets, and the d-faces are called *cells*. The (combinatorial) complexity of the arrangement is defined as the sum of the number of i-faces, for $0 \leqslant i \leqslant d$. It is well known that the complexity of an arrangement of n hyperplanes in d-dimensional space is $O(n^d)$. Moreover, it is possible to construct the arrangement—that is, compute all its faces and the incidences between the faces—in $O(n^d)$ time. All this can be found in Edelsbrunner's textbook [50], which is a general source of information on algorithmic and combinatorial aspects of arrangements. As with other complex objects, it is often convenient to *triangulate* an arrangement

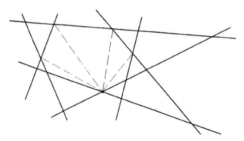

Figure 2.2: A (a part of) two-dimensional arrangement and the bottom vertex triangulation for one of its cells.

$\mathcal{A}(H)$, that is, to subdivide its cells into d-dimensional simplices with disjoint interiors. A simple way of doing this, due to Clarkson [35], is to construct the *bottom vertex triangulation* of the arrangement (see Figure 2.2), which is defined as follows. Consider a cell of $\mathcal{A}(H)$. This cell is a convex polytope in $\mathbb{E}^d$. It can be triangulated as follows. If $d = 1$ then the polytope—let's call it $\mathcal{P}$—is an interval on the real line and, hence, it already is a simplex. Otherwise, fix a vertex v of $\mathcal{P}$, for example the one with smallest x_d-coordinate, the so-called bottom vertex. Next, triangulate in a recursive manner each of the facets of $\mathcal{P}$—which are convex polytopes in $\mathbb{E}^{d-1}$—that are not incident to v. Finally, extend each $(d-1)$-dimensional simplex $\tilde{s}$ thus obtained into a d-dimensional simplex s by adding v as a vertex. More precisely, s is the convex hull of $\tilde{s}$ and v. We call the resulting triangulation the bottom vertex triangulation of c. By triangulating every cell of $\mathcal{A}(H)$ in this way we obtain the bottom vertex triangulation of $\mathcal{A}(H)$. (For the unbounded cells of the arrangement, this is not well defined. The definition is easily extended to handle unbounded cells correctly [35].) After the arrangement has been constructed it is easy to compute the simplices of the bottom vertex triangulation—whose total number is linear in the complexity of the arrangement—in time $O(n^d)$.

One last piece of terminology. We say that an object o in $\mathbb{E}^d$ is *above (below)* an object o' if there is a line parallel to the x_d-axis that intersects o in a point q and o' in a point q' such that $q_d > q'_d$ ($q_d < q'_d$).

2.3 Segment Trees

A segment tree is a data structure that is used to store a set of intervals on the real line. The structure allows us to select all intervals that contain a query point in terms of a logarithmic number of so-called canonical subsets. Since their introduction by Bentley [13] in 1977, segment trees have found extensive use in computational geometry; they are at the basis of many structures that store axis-parallel or c–oriented objects. See, for example, the textbooks by Preparata and Shamos [104] and Mehlhorn [85]. Segment trees will also be used at several places in this book.

Let $S = \{I_1, \ldots, I_n\}$ be a set of n possibly overlapping half-open[1] intervals on the real line, and let $I_i = [x_i : x'_i)$. The $m \leqslant 2n$ different endpoints of the intervals in S partition the real line into $m + 1$ half-open *elementary intervals*. The segment tree $\mathcal{T}$ that stores S is a balanced binary tree with $m+1$ leaves. The elementary intervals are associated in an ordered way with these leaves: the leftmost leaf corresponds to the leftmost interval, and so on. Each of the internal nodes has an interval associated with it that is the union of the intervals associated with its two children. In other words, each node ν of $\mathcal{T}$ has associated with it an interval that is the union of the elementary intervals of the leaves in the subtree rooted at ν. We denote this interval by I_ν. An interval $I_j \in S$ is stored at those nodes ν such that $I_\nu \subseteq I_j$, but $I_{parent(\nu)} \not\subseteq I_j$. One easily checks that I_j is precisely the disjoint union of the intervals I_ν over all nodes ν where I_j is stored. Moreover, I_j can be stored with at most two nodes at every level of $\mathcal{T}$: for any node ν in between these two nodes $I_{parent(\nu)} \subseteq I_j$ must hold, and therefore I_j will not be stored at ν. See Figure 2.3. We denote the subset of intervals that are stored at some node ν by $S(\nu)$. We refer to $S(\nu)$ as the *canonical subset of node ν*.

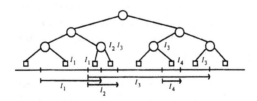

Figure 2.3: A segment tree storing four segments.

---

[1] The structure is easily adapted so that it can handle open and closed intervals as well.

The search path in $\mathcal{T}$ of a real number x visits exactly those nodes ν such that $x \in I_\nu$. Recall that the interval corresponding to a node is the disjoint union of the intervals corresponding to its two children, and note that $I_{root(\mathcal{T})}$ is the real line. Therefore the nodes that are visited form a path down the tree, starting at the root and ending at the leaf whose elementary interval contains x.

Theorem 2.1 *A segment tree $\mathcal{T}$ for a set S of n intervals uses $O(n \log n)$ storage and can be built in $O(n \log n)$ time. The subset of intervals in S that contain a query point x is the disjoint union of the canonical subsets $S(\nu)$ of all nodes ν on the search path of x in $\mathcal{T}$.*

Proof: Each interval in S is stored at most $O(\log n)$ times, since $\mathcal{T}$ is balanced and each interval I_j can be stored with at most two nodes at every level of $\mathcal{T}$. This proves the bound on the amount of storage. The skeleton of $\mathcal{T}$ (the balanced binary tree with $m + 1$ leaves) can be built in $O(n)$ time after sorting the endpoints of the intervals in $O(n \log n)$ time. Moreover, locating the nodes where an interval has to be stored can be done in $O(\log n)$ time per interval. See, for example, Preparata and Shamos [104].

An interval I_j is only stored at nodes ν such that $I_\nu \subseteq I_j$, and the search path of a point x only visits nodes with $x \in I_\nu$. It follows that the intervals stored in the sets $S(\nu)$ for nodes ν on the search path of x must contain x. Moreover, an interval is never stored at two nodes that are on a common path in $\mathcal{T}$, because then the property $I_{parent(\nu)} \nsubseteq I_j$ would be violated for the lowest of these two nodes. Finally, each interval in S containing x must contain the elementary interval of the leaf where the search ends. Hence, it is either stored at that leaf or at an ancestor. It follows that the set of intervals containing x is exactly the disjoint union of the sets $S(\nu)$ over all nodes ν on the search path of x. $\square$

The way the canonical subsets $S(\nu)$ are stored depends on the application. For example, if we want to know the number of intervals that contain a query point, then we only need to store at each node ν the cardinality of its canonical subset. Note that this reduces the amount of storage to $O(n)$. In other cases the associated structures that store the canonical subsets are more complex data structures, giving rise to a *multi-level data structure*. This is especially the case if the segment tree is used to solve two- or more-dimensional problems.

To illustrate the concept of multi-dimensional segment trees, let us consider the problem of storing a set of axis-parallel rectangles in the plane such that we can select all rectangles that contain a query point $q = (q_x, q_y)$. The solution we sketch was reported by Edelsbrunner and Maurer [52].

An axis-parallel rectangle is the Cartesian product of an x-interval and a y-interval. We build a segment tree $\mathcal{T}$ for the set of x-intervals of the rectangles. A node ν in $\mathcal{T}$ has a certain x-interval I_ν associated with it. Since $\mathcal{T}$ is used to store two-dimensional objects, ν can be considered to represent the region $I_\nu \times [-\infty : \infty]$ in the plane; we call this region the *slab* corresponding to ν. Using the same terminology, we can say that the canonical subset of a node ν contains those rectangles that span the slab corresponding to ν, but not the slab corresponding to $parent(\nu)$. When we consider a

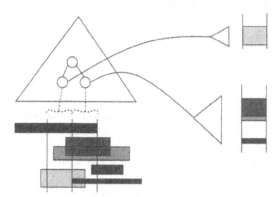

Figure 2.4: A two-dimensional segment tree.

rectangle in a canonical subset $S(\nu)$, we always restrict our attention to the part of the rectangle inside the slab corresponding to ν. A search with q_x in $\mathcal{T}$ gives us $O(\log n)$ canonical subsets $S(\nu)$ whose union is the set of rectangles whose x-interval contains q_x. Of these rectangles we are interested in the ones whose y-interval contains y. Thus, each canonical subset $S(\nu)$ is stored in a segment tree $\mathcal{T}_\nu$ on the y–intervals of the rectangles. See Figure 2.4. To select the rectangles containing q we first search with q_x in $\mathcal{T}$ and then we search with q_y in the trees $\mathcal{T}_\nu$ for nodes ν on the search path. In each $\mathcal{T}_\nu$ we find $O(\log n)$ nodes; hence, we find $O(\log^2 n)$ nodes in total, whose canonical subsets together contain the rectangles containing q. The total amount of storage this two-dimensional segment tree uses is $\sum_{\nu \in \mathcal{T}} O(|S(\nu)| \log |S(\nu)|) = O(n \log^2 n)$.

Finally, we discuss the dynamization of segment trees. Consider the insertion of an interval into an ordinary one-level segment tree. It is fairly straightforward to insert the interval if its endpoints are already present. When this is not the case, however, things get a little more complicated. The problem is that a new endpoint causes an elementary interval to be split. Thus a leaf has to be replaced by an internal node with two children, and the tree might become unbalanced. Fortunately, these problems can be resolved using the general technique of Willard and Lueker [123]. This technique even allows us to perform updates in multi-level segment trees.

The following theorem summarizes the results on multi-level segment trees and their dynamization.

Theorem 2.2 *Suppose that the canonical subsets $S(\nu)$ in a segment tree are stored in associated structures that use $M_A(|S(\nu)|)$ storage and have $U_A(|S(\nu)|)$ update time. Then the total amount of storage used by the segment tree and its associated structures is $O(M_A(n) \log n)$, and the update time is $O(U_A(n) \log n)$.*

2.4   Fractional Cascading

Fractional cascading is a general technique due to Chazelle and Guibas [25, 26] that can be used to speed up the query time of a large class of data structures. We discuss this technique on the basis of the segment trees that we introduced in the previous section.

Suppose we want to preprocess a set S of n axis-parallel rectangles in the plane such that we can decide if a query point q is contained in the union of the rectangles. A possible way of solving this would be to compute the union explicitly and preprocess it for point location queries. However, the union may have quadratic complexity so this solution uses a large amount of storage. A better solution can be obtained using segment trees.

Store the rectangles in a segment tree $\mathcal{T}$ on their x-intervals, and store at each node ν in $\mathcal{T}$ the union of the y-intervals of the rectangles in $S(\nu)$. Note that the union of a set of intervals again consists of a number of intervals. At node ν the y-coordinates of the endpoints of the intervals of the union are stored in a sorted list $\mathcal{L}_\nu$. Clearly, the complexity of the union and, hence, the size of $\mathcal{L}_\nu$ is $O(|S(\nu)|)$. By Theorem 2.2 the total amount of storage used by the data structure is $O(n \log n)$. Notice that a query point q is contained in the union of the rectangles if and only if q_y is contained in one or more of the unions of the y-intervals stored at nodes ν on the search path of q_x in $\mathcal{T}$. Thus to answer a query we have to search with q_y in the lists $\mathcal{L}_\nu$ of each node ν on the search path. Because these lists are ordered we can do a binary search on them, leading to an overall query time of $O(\log^2 n)$. But using fractional cascading

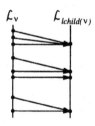

Figure 2.5: The pointers needed for fractional cascading.

we can do better. The basic observation behind this technique is that the position of q_y in one of the lists can give us information about the position of q_y in the lists of its children. For example, if the y-coordinates in $\mathcal{L}_{lchild(\nu)}$ form a subset of the y-coordinates in $\mathcal{L}_\nu$, then the position of q_y in $\mathcal{L}_\nu$ uniquely determines the position in $\mathcal{L}_{lchild(\nu)}$. So we add a pointer from each element in $\mathcal{L}_\nu$ to the smallest element which is at least as large in the list $\mathcal{L}_{lchild(\nu)}$. Given the position of q_y in $\mathcal{L}_\nu$—that is, knowing the smallest element in $\mathcal{L}_\nu$ which is at least as large as q_y—we can now determine the position of q_y in $\mathcal{L}_{lchild(\nu)}$ by following a pointer. See Figure 2.5. Of course, we are rarely in the fortunate situation sketched above; the lists that we have to search usually have little in common. This problem can be overcome by copying certain elements of the lists into other lists. This has to be done in a clever way, since we do not want to increase the amount of storage too much. Chazelle and Guibas [25] have shown that the copying can be done in such a way that the search in the list of some node ν can be done in constant time if we already know the position of the query value in the list of $parent(\nu)$. Moreover, their scheme does not affect the asymptotic preprocessing time or the storage bound.

Returning to our example, we see that we can answer queries in $O(\log n)$ time: we perform a binary search in the list stored at the root of $\mathcal{T}$, and then we can do the searches in the lists stored at the remaining nodes on the search path in constant time per list.

Fractional cascading can also be applied to other structures where we have to search a number of lists with the same value. Roughly speaking, if the lists are associated with nodes in an undirected, connected, acyclic graph of bounded degree, then fractional cascading can be used to reduce the time needed to search the lists associated with the nodes on a path in the graph. We refer the reader to [25] for further details. To avoid introducing unnecessary notation, we will not state the result on fractional cascading in its full generality, but we only state it the way we will use it in the sequel.

Theorem 2.3 (Chazelle and Guibas [25]) *Let $\mathcal{T}$ be a binary tree with n nodes, and suppose that each node ν in $\mathcal{T}$ stores an ordered list $\mathcal{L}_\nu$. Fractional cascading allows us to compute the position of a query value in $\mathcal{L}_\nu$ in constant time if we know the position of the query value in $\mathcal{L}_{parent(\nu)}$. The time needed to set up the fractional cascading structure is $O(n + \sum_{\nu \in \mathcal{T}} |\mathcal{L}_\nu|)$.*

To get the search started, we always perform—although we usually do not explicitly say so—a binary search in the list which is stored at the root of the tree. This way the total time needed to search the $O(\log n)$ lists on a search path in a balanced binary tree is bounded by $O(\log n)$.

Mehlhorn and Näher [86] have shown that it is also possible to use fractional cascading in a dynamic environment. We will mainly be interested in dynamic fractional cascading in connection with segment trees. Suppose we have a set of segments in the plane that are parallel to the x-axis. Given a query point q we want to find the segment directly below q. In other words, we are looking for the segment $s = [s_x : s'_x] \times s_y$ with $q_x \in [s_x : s'_x]$, $q_y \geqslant s_y$ and s_y maximal. This problem can be solved by building a segment tree $\mathcal{T}$ on the x-intervals of the segments in S, and storing at each node ν the segments in $S(\nu)$ in a list $\mathcal{L}_\nu$ that is sorted on y-coordinate. In a static setting we can apply fractional cascading to obtain $O(\log n)$ search time. Note that the pointers that we have in the fractional cascading structure need to be changed if we add new segments to the structure. Mehlhorn and Näher have shown that it is possible to make the necessary changes in $O(\log n \log \log n)$ amortized time. The search time in the dynamic setting increases to $O(\log n \log \log n)$.

Theorem 2.4 (Mehlhorn and Näher [86]) *A segment tree $\mathcal{T}$ for n segments, where the sets $S(\nu)$ are stored in ordered lists, can be maintained in $O(\log n \log \log n)$ amortized time per update. The structure uses $O(n \log n)$ storage. Locating a query value in the lists on a search path in $\mathcal{T}$ takes $O(\log n \log \log n)$ time in total.*

It should be noted that the result of Mehlhorn and Näher is more general than stated here: it applies to the same kind of structures for which the static result can be used. See [86] for details.

2.5   Cuttings and Random Sampling

Cuttings have become a basic tool for solving problems concerning sets of hyperplanes in higher dimensions. In this section we define cuttings, and we show how to construct and use them. We do this on the basis of point location in arrangements of hyperplanes.

Let H be a set of n hyperplanes in $\mathbb{E}^d$. Recall that $\mathcal{A}(H)$, the arrangement of H, is the subdivision of $\mathbb{E}^d$ into connected pieces of various dimensions that is induced by H. We want to preprocess $\mathcal{A}(H)$ for point location, that is, we want to store $\mathcal{A}(H)$ in a data structure such that we can efficiently determine the cell of $\mathcal{A}(H)$ containing a query point. In other words, we want to determine the position of the query point relative to each of the hyperplanes.

If $d = 1$ then the hyperplanes are points on the real line and the point location problem is readily solved with a balanced binary search tree on the points. This is due to the fact that the real line can be partitioned into two intervals such that each interval contains approximately half of the points. More generally, the real line is easily partitioned into $O(r)$ intervals such that each interval contains at most n/r of the points, leading to a balanced search tree of branching degree $O(r)$. Clarkson [34, 35] was the first to realize that this divide-and-conquer approach can be extended to higher dimensions by partitioning $\mathbb{E}^d$ into simplices such that the interior of each simplex is cut by at most n/r of the hyperplanes in H. Such a set of simplices is called a $(1/r)$-*cutting* for H, denoted by $\Xi(H)$. The number of simplices that are used is the *size* of the cutting. See 2.6 for an illustration of the concept. We leave the description of the point location structure based on cuttings for later, and turn our attention to the existence and construction of cuttings of small size. In dimensions greater than one it is generally not possible to find a $(1/r)$-cutting of size $O(r)$: a simple arrangement of hyperplanes in $\mathbb{E}^d$ has $\Theta(n^d)$ vertices, so we need $\Omega(r^d)$ simplices if we want each simplex to intersect no more than n/r hyperplanes. As we shall see below, this number is not only necessary but also sufficient.

2.5.1   Random Sampling

Random sampling is an easy and efficient way to construct cuttings. The idea is that a randomly chosen subset of a set is likely to be a 'good approximation' of that set. Thus, we hope that if we take a random subset $R \subset H$, then each cell in the arrangement $\mathcal{A}(R)$ is intersected by few hyperplanes in H; after all, (the interior of)

Figure 2.6: A $(1/2)$-cutting of size ten for a set of six lines in the plane.

the cell is intersected by none of the hyperplanes in R. This simple idea does not quite work yet. Consider a set H such that there is a cell in $\mathcal{A}(H)$ that is bounded by all hyperplanes. In this case there will be a cell in $\mathcal{A}(R)$ that is bounded by all hyperplanes in R. Moreover, this cell is intersected by all of the remaining hyperplanes in $H - R$, which is clearly too much. Fortunately, it takes only a slight change to make the idea work: triangulate the cells of $\mathcal{A}(R)$, for example using the bottom vertex triangulation described in Section 2.2. This gets rid of the complex cells in $\mathcal{A}(R)$ and ensures the desired property, as has been proved by Clarkson [34]. In the following lemma, a random sample $R \subset H$ of size r is a subset of H obtained by picking r hyperplanes from H at random, with replacement; thus, at every pick each hyperplane has probability $1/n$ that it is chosen.

Theorem 2.5 (Clarkson [34]) *Let H be a set of n hyperplanes in $\mathbb{E}^d$, and let $R \subset H$ be a random sample of size r. The probability that there is a simplex in the triangulated arrangement $\mathcal{A}(R)$ that is intersected by more than αn hyperplanes in H is $O(r^{d(d+1)}(1-\alpha)^{r-d(d+1)})$. For any fixed i there is a suitable $\alpha = O(\log r/r)$ such that this probability is at most $1/r^i$.*

Stated differently, Theorem 2.5 says that the triangulated arrangement $\mathcal{A}(R)$ of a random subset $R \subset H$ is an $O(\log r/r)$-cutting for H of size $O(r^d)$, with probability at least $1 - \frac{1}{r^i}$. Note that this immediately proves the existence of such a cutting, and also suggests a simple algorithm for constructing it: Take a random sample and test if it has the required property; if not, take another random sample, and so on, until a good sample is found. We expect to succeed within a constant number of trials, so the algorithm takes $O(nr^d)$ randomized time.

2.5.2 Deterministic Computation of Cuttings

The random sampling method for constructing cuttings has two disadvantages. First, its preprocessing is not deterministic. Second, it has an extra $O(\log r)$ factor that does not appear in the lower bound. Therefore, after the pioneering work of Clarkson much research has been devoted to finding new ways to construct cuttings [2, 18, 23, 77, 78, 79]. Chazelle and Friedman [23] were the first to show that a $(1/r)$-cutting of size

$O(r^d)$ always exists. They also gave an efficient randomized procedure for computing such a cutting. Later, Matoušek [79] gave an efficient deterministic algorithm that can compute, for almost all r, a $(1/r)$-cutting of (asymptotically) optimal size. Recently, Chazelle succeeded in devising an algorithm that computes an optimal cutting for all values of r.

Theorem 2.6 (Chazelle [18]) *Given a set H of n hyperplanes in $\mathbb{E}^d$ and a parameter $r \leqslant n$, it is possible to compute a $(1/r)$-cutting of size $O(r^d)$ for H, in time $O(nr^{d-1})$.*

This result is in theory stronger than the result of Clarkson, so we will use Theorem 2.6 instead of Theorem 2.5 most of the time. In practice the simplicity of Clarkson's method is probably more important than the fact that Chazelle's method is deterministic and produces an optimal result. Indeed, the extra $O(\log r)$ factor usually does not influence the final results. Besides its simplicity, the random sampling method has the advantage that the resulting cutting is somewhat more structured, since its simplices are the simplices of a triangulated arrangement. In some applications this is crucial and, hence, Chazelle's deterministic method cannot be used.

2.5.3 Using Cuttings

Cuttings can be used to solve a variety of problems concerning sets of hyperplanes. Let us illustrate this by giving a simple structure for point location in the arrangement $\mathcal{A}(H)$ of a set H of hyperplanes in $\mathbb{E}^d$. Each cell in the arrangement has a unique label. The answer to a point location query with point q is the label of the cell containing q. For the sake of exposition, we assume that q is not contained in any of the hyperplanes, which makes the answer unique. The method is due to Clarkson [34], with the random sampling technique he uses replaced by Chazelle's deterministic cuttings. The structure is a tree of branching degree $O(r^d)$ whose root stores a $(1/r)$-cutting $\Xi(H)$ for the set H of hyperplanes. Each child ν of the root corresponds to a simplex s_ν of $\Xi(H)$. The simplex s_ν partitions H into three subsets: a subset $H^+(\nu)$ of hyperplanes that are above s_ν, a subset $H^-(\nu)$ of hyperplanes that are below s_ν, and a subset $H^\times(\nu)$ of hyperplanes that intersect s_ν. Node ν is the root of a recursively defined structure on the set $H^\times(\nu)$. If the number of hyperplanes at some node ν in the tree drops below some constant (we call ν a *small node*), we just store the complete arrangement. That is, every cell of the arrangement $\mathcal{A}(H^\times(\nu))$ is stored in a separate leaf that is a child of the small node.

A search with point q in this structure is performed as follows. First we determine which of the $O(r^d)$ simplices of $\Xi(H)$ contains q. Let s_ν be this simplex. Now we know the position of q relative to all hyperplanes except the $O(n/r)$ hyperplanes in $H^\times(\nu)$. Thus we recursively search in the subtree rooted at ν, until we reach a small node. To determine the position of q relative to the remaining hyperplanes, we simply check each cell of the arrangement that corresponds to the small node. This cell corresponds to a leaf below the small node. It is readily seen that the leaf where the search ends uniquely determines the cell of $\mathcal{A}(H)$ that contains q. To be able to answer a point location query, we just have to store during the preprocessing phase with each leaf the label of corresponding cell of $\mathcal{A}(H)$.

When we analyze the performance of this data structure, we see that the preprocessing time $T(n)$ satisfies[2]

$$T(n) \;=\; O(nr^{d-1}) + O(r^d)T(n/r). \tag{2.1}$$

The query time $Q(n)$ satisfies

$$Q(n) = O(r^d) + Q(n/r).$$

It is easily seen that the query time is $O(\log n)$ if we choose r a constant greater
than 1. Moreover, for any $\varepsilon > 0$ we can obtain a bound of $T(n) = O(n^{d+\varepsilon})$ on the
preprocessing time by choosing r large enough, as follows from the First Recurrence
Lemma that is given below. We will use this general lemma repeatedly in Part B to
analyze the preprocessing time and the amount of storage of our ray shooting data
structures. Before we state the lemma, let us summarize our findings on point location
in arrangements.

Theorem 2.7 (Clarkson [34]) *Let H be a set of n hyperplanes in $\mathbb{E}^d$. For any $\varepsilon > 0$,
there exists a structure that uses $O(n^{d+\varepsilon})$ storage and preprocessing time, and answers
point location queries in the arrangement $\mathcal{A}(H)$ in $O(\log n)$ time.*

(In fact, better bounds can be obtained for point location; see Remark 2.11 at the
end of this section.) We have promised the reader a general lemma to analyze the
preprocessing time and amount of storage of data structures that are based on cuttings. The lemma can both be used to analyze one-level data structures—such as the
point location structure described above—and to analyze multi-level data structures.
Before we state the lemma, we give an example of a two-level data structure that is
based on cuttings and can be analyzed with the First Recurrence Lemma. The key
observation is that the point location structure described above in fact gives us the
hyperplanes lying above a query point in a small number of *canonical subsets*. These
canonical subsets are the sets $H^+(\nu)$ for nodes ν on the search path. This observation enables us to construct multi-level data structures that solve more complicated
problems than point location. Consider, for example, the problem where we want to
count all hyperplanes that lie above a query line segment. Observe that a hyperplane
lies above a segment if and only if it lies above both endpoints. We build a tree $\mathcal{T}$ as
described above on the set H of hyperplanes. At each node ν in $\mathcal{T}$, we store the set
$H^+(\nu)$ in an associated structure $\mathcal{T}_\nu$; this associated structure is itself also a structure
as described above. At nodes μ in $\mathcal{T}_\nu$ we store the cardinality of the canonical subset
$H^+(\mu)$. A query searches the main tree with one of the endpoints of the segment.
For each node ν on the search path, we search in the associated structure with the
other endpoint. The sets $H^+(\mu)$ for nodes μ on the search paths in the associated
structures contain exactly the hyperplanes lying above both endpoints, that is, lying
completely above the segment. Hence, the total number of such hyperplanes can be
computed by adding up the cardinalities of these canonical subsets. If we choose r to

---

[2]For this and all other recurrences given in the sequel, we omit the boundary conditions. All
functions for which we give recurrences have constant value if n is constant, and the precise boundary
conditions do not influence the asymptotic behavior of the functions.

be a constant, then the depth of the main tree is $O(\log n)$, leading to $O(\log^2 n)$ query time in total. The preprocessing time and the amount of storage $T(n)$ for the new structure satisfies

$$T(n) = O(r^d n^{d+\varepsilon}) + O(r^d)T(n/r),$$

because the associated structure stored for each of the $O(r^d)$ children uses $O(n^{d+\varepsilon})$ preprocessing. The following lemma gives a solution to this recurrence, if r is sufficiently large (the condition on r is made precise in the proof of the lemma).

Lemma 2.8 [First Recurrence Lemma] *Let* $T(n) = O(r^d n^{d+\varepsilon}) + O(r^d)T(n/r)$, *where* $\varepsilon > 0$ *and* d *are constants and* r *is a sufficiently large, not necessarily constant, parameter. Then* $T(n) = O(r^d n^{d+\varepsilon})$.

Proof: Let c be a constant such that $T(n) \leqslant c r^d n^{d+\varepsilon} + c r^d T(n/r)$ for all n, and let $c' > 2c$ be a constant such that $T(n) \leqslant c' r^d n^{d+\varepsilon}$ for small n. For $r > (2c)^{1/\varepsilon}, r \geqslant 2$ we can then show by induction that $T(n) \leqslant c' r^d n^{d+\varepsilon}$ for all n:

$$
\begin{aligned}
T(n) &\leqslant c r^d n^{d+\varepsilon} + c r^d T(\tfrac{n}{r}) \\
&\leqslant c r^d n^{d+\varepsilon} + c r^d (c' r^d (\tfrac{n}{r})^{d+\varepsilon}) \\
&= c' r^d n^{d+\varepsilon} (\tfrac{c}{c'} + \tfrac{c}{r^\varepsilon}) \\
&\leqslant c' r^d n^{d+\varepsilon}
\end{aligned}
$$

$\square$

It is worth noting that the first term of the recurrence is $O(r^d n^{d+\varepsilon})$ and that this is also the solution to the recurrence. In terms of data structures, this means that the associated structures use $O(n^{d+\varepsilon})$ storage, and that the amount of storage used by the total structure is $O(r^d n^{d+\varepsilon})$; for constant r these two bounds are asymptotically the same. Hence, we can build multi-level data structures without any (asymptotic) costs involved for the preprocessing time and the amount of storage.

A second interesting observation that we can make is the following. The first term of recurrence (2.1), which gives the preprocessing time and the amount of storage for a single-level structure, was only $O(nr^{d-1})$. This implies that for *any* $\varepsilon > 0$ we can choose r such that the solution is $O(n^{d+\varepsilon})$. For a two-level data structure such as the one described above, there are two parameters r_1 and r_2: a parameter r_1 for the main tree and a parameter r_2 for the associated trees. The associated trees are one-level structures, so for any $\varepsilon > 0$ we can obtain a structure that uses $O(n^{d+\varepsilon})$ storage by choosing r_2 large enough. Now consider the total two-level structure. Using the First Recurrence Lemma once more, we see that for any $\varepsilon > 0$ we can obtain $O(n^{d+\varepsilon})$ storage and preprocessing time, by choosing both r_1 and r_2 large enough.

A final observation is that the First Recurrence Lemma is also valid if r is a function of n. Therefore we can play the following trick. Consider the two-level structure. For the associated trees we take the parameter r_2 to be a constant, as before. For the main tree, however, we set $r_1 = n^{\varepsilon_1}$, for some constant $\varepsilon_1 > 0$. By choosing r_2 large enough we can obtain, for any $\varepsilon > 0$, a bound on the total amount of storage of $O((r_1)^d n^{d+\varepsilon}) = O(n^{d+d\varepsilon_1+\varepsilon})$. Note that we still have the freedom to choose ε_1 as small as we like. Hence, for any $\varepsilon_2 > 0$ we can build a structure that uses $O(n^{d+\varepsilon_2})$

storage. So choosing $r_1 = n^{\varepsilon_1}$, instead of choosing r_1 a constant, does not increase the bound on the storage—at least, nobody will notice. But why would we want to have $r_1 = n^{\varepsilon_1}$ anyway? To see this, consider a query in the two-level structure for finding all hyperplanes above two query points. A query in this structure searches in the cutting $\Xi(H)$ stored at the root, to determine the simplex s_ν containing the first query point. Then the associated structure $\mathcal{T}_\nu$ is queried with the second query point, taking $O(\log n)$ time. Finally, the query algorithm recursively searches the subtree rooted at ν. If we search for the simplex s_ν that contains the first query point in a brute-force way, then the query time $Q(n)$ satisfies $Q(n) = O((r_1)^d) + O(\log n) + Q(n/r_1)$, leading to $Q(n) = O(\log^2 n)$ if r_1 is a constant. But if we use a fast point location structure to find this simplex in $O(\log n)$ time, then the recurrence becomes

$$Q(n) = O(\log n) + Q(n/r_1).$$

Our Second Recurrence Lemma shows that the total query time remains $O(\log n)$ if we choose $r_1 = n^{\varepsilon_1}$.

Lemma 2.9 [Second Recurrence Lemma] *Let $Q(n) = O(\log n) + Q(n/r)$, where $r = n^\varepsilon$ and $\varepsilon > 0$ is a constant. Then $Q(n) = O(\log n)$.*

Proof: Let c be a constant such that $Q(n) \leqslant c \log n + Q(n/r)$ for all n, then we have:

$$
\begin{aligned}
Q(n) &\leqslant c \log n + Q(\tfrac{n}{r}) \\
&= c \log n + Q(n^{1-\varepsilon}) \\
&\leqslant c \log n + c \log n^{1-\varepsilon} + Q(n^{(1-\varepsilon)^2}) \\
&= c[1 + (1-\varepsilon)] \log n + Q(n^{(1-\varepsilon)^2}) \\
&\leqslant c \sum_{i=0}^{\infty} (1-\varepsilon)^i \log n \\
&= O(\log n)
\end{aligned}
$$

$\square$

Following the same scheme, we can construct structures with three or more levels that have $O(\log n)$ query time and use $O(n^{d+\varepsilon})$ storage and preprocessing time: for all the new levels that we add, we choose the parameter r to be n^δ, for a sufficiently small $\delta > 0$. For example, for any $\varepsilon > 0$ there exists a structure to count the number of hyperplanes lying above a set of $d+1$ points with $O(\log n)$ query time that uses $O(n^{d+\varepsilon})$ storage and preprocessing time. (This problem is better known in its dual setting, where it is called the *simplex range searching problem*. In this setting we are given a set of points and we want to count the number points inside a query simplex.)

Remark 2.10 To keep the query time down to $O(\log n)$ we need a fast point location structure for cuttings. In general this can be a hard problem, but the fact that the size of the cuttings is only $O(n^{d\delta})$ saves us. See Section 7.2.3 for the trick that can be used in that case.

Remark 2.11 The result for point location in an arrangement of hyperplanes in $\mathbb{E}^d$ that is given in this section is not the best possible. Chazelle and Friedman [24] have shown that one can shave off the $O(n^\varepsilon)$ factor from the storage bound, leading to a structure that uses $O(n^d)$ storage, with $O(\log n)$ query time. The preprocessing time of their solution is quite high, however. A simpler solution with $O(n^d)$ storage and preprocessing time can be obtained if we use some special property of Chazelle's cuttings [18], namely that he constructs the cuttings by refining subsequent cuttings. See [18] for details.

2.6 Partition Trees

Consider the following problem. We want to store a given set of points in $\mathbb{E}^d$ in a data structure, such that we can count the number of points above a query hyperplane quickly. For readers who are familiar with dualization—which is discussed in the next section—it is clear that this problem is equivalent to the problem of counting the number of hyperplanes in a given set that are above a query point. In the previous section we have seen that cuttings can be used to answer this question in $O(\log n)$ time with a structure that uses $O(n^{d+\varepsilon})$ storage. But what if we do not have this amount of storage available and want to use, say, only a roughly linear amount of storage? In this section we show that we can obtain $O(n^{1-1/d+\varepsilon})$ query time with a structure—usually called a *partition tree*—that uses a near-linear amount of storage. By combining this in a clever way with the results of the previous section one can obtain a whole range of trade-offs between query time and amount of storage: for any $n \leqslant m \leqslant n^d$, it is possible to achieve $O(n^{1+\varepsilon}/m^{1/d})$ query time with a structure that uses $O(m^{1+\varepsilon})$ storage. The same efficiency can be obtained for the simplex range searching problem. In view of Chazelle's lower bounds [16], which state that the query time of a structure for simplex range searching that uses $O(m)$ storage must be $\Omega(n/(m^{1/d}\log n))$, this result is near-optimal.

2.6.1 Linear Space Partition Trees

Let S be a set of n points in $\mathbb{E}^d$, and let h be a (query) hyperplane. Partition trees are based on an idea similar to cuttings, namely subdividing space into simplices such that we can use a divide-and-conquer approach. In order for this approach to be efficient, we have make sure that any query hyperplane intersects only 'few' simplices. In fact, what is important is not so much the number of simplices that are intersected, but the total number of points in them. The first partition tree based on this principle was described by Willard [122]. He obtained $O(n^{0.774})$ query time using linear space for the planar case. This result was subsequently improved and generalized to higher dimensions by a number of people [28, 29, 55, 70, 77, 80]. We describe the recent result of Matoušek [80], which is the best known result so far.

A *simplicial partition* for a set S of n points in $\mathbb{E}^d$ is defined to be a collection $\Psi(S) = \{(S_1, s_1), \ldots, (S_m, s_m)\}$, where the S_i's are disjoint subsets of S whose union is S, and s_i is a simplex containing S_i. Observe that s_i may contain other points as well. Thus the simplices are not necessarily disjoint, as is the case for cuttings. We call m, the number of simplices, the *size* of $\Psi(S)$. See Figure 2.7 for a two-dimensional illustration. In this figure, the points that are in the same subset S_i have the same pattern. The *crossing number* of a hyperplane h with respect to $\Psi(S)$ is the number

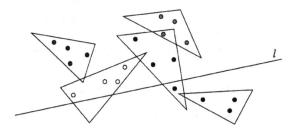

Figure 2.7: A simplicial partition of size five for a set of points in the plane.

of simplices of $\Psi(S)$ that are intersected by h. Thus the crossing number of the line l in Figure 2.7 is two. The crossing number of $\Psi(S)$ is the maximum crossing number over all hyperplanes h. Finally, we define the *relative zone size* of h to be $\frac{1}{n}\sum_{s_i \cap h \neq \varnothing} |S_i|$, that is, it is the fraction of points in S that belong to simplices that are intersected by h. In Figure 2.7, for example, the relative zone of line l is 8/19. The relative zone size of $\Psi(S)$ is the maximum relative zone size over all hyperplanes h. Notice that the points in the simplices crossed by h are the only points of which we do not know their position relative to h: the sets S_i corresponding to simplices that are not intersected by h lie completely above or completely below h. Thus only the sets S_i whose corresponding simplex is intersected need recursive treatment. Hence, the relative zone size of a simplicial partitioning determines its efficiency. The following theorem, proved by Matoušek [80], shows that good simplicial partitions exist and can be computed efficiently. These simplicial partitions have the extra property that the points in S are distributed nicely among the sets $S_1, \ldots, S_m$. More precisely, the partitionings are *fine*, that is, $|S_i| \leqslant 2n/m$ for every $1 \leqslant i \leqslant m$.

Theorem 2.12 (Matoušek [80]) *Let S be a set of n points in $\mathbb{E}^d$ and $r \leqslant n$ a parameter. Then, for any $\varepsilon > 0$, it is possible to construct a fine simplicial partition of size r for S with crossing number $O(r^{1-1/d})$, in time $O(n^{1+\varepsilon})$.*

In this theorem, the constant in the bound on the crossing number depends on ε. Note that the fact that the partition is fine, together with the bound on the crossing number, implies that the relative zone size of the partition is $O(r^{-1/d})$.

The data structures that result from simplicial partitionings have the same flavor as the data structures resulting from cuttings. For example, if we want to count the number of points above a query hyperplane, then we proceed as follows. We compute a fine simplicial partition $\Psi(S)$ of size $O(r)$ for S with crossing number $O(r^{1-1/d})$ and such that $|S_i| \leqslant n/r$, where r is a (large enough) constant. The data structure is a tree $\mathcal{T}$ of branching degree $O(r)$, with a one-to-one correspondence between the simplices of $\Psi(S)$ and the children of $root(\mathcal{T})$. The child ν_{s_i} corresponding to simplex

s_i is the root of a recursively defined structure for the set S_i. We call S_i the *canonical subset of* ν_{s_i}. We also store for each simplex s_i the cardinality of S_i. A query with a hyperplane h in this structure is performed as follows. We check for each simplex of $\Psi(S)$ whether it intersects h. For each intersected simplex s_i, we count the number of points in S_i that are above h by recursively searching in the subtree rooted at ν_{s_i}. The other sets S_i are either fully contained in h^-, in which case they do not contribute to the total count, or they are fully contained in h^+. In the latter case the contribution is just the cardinality of S_i, which we have precomputed. Thus the query time $Q(n)$ satisfies the recurrence

$$Q(n) = O(r) + O(r^{1-1/d})Q(n/r).$$

A close look at the First Recurrence Lemma shows that it also applies to this recurrence (if $d \geqslant 2$). Hence, for any $\varepsilon > 0$, we can obtain a query time of $Q(n) = O(n^{1-1/d+\varepsilon})$ by choosing r large enough. The amount of storage that we have used is linear, since we only store the cardinalities of the sets S_i at a node.

Theorem 2.13 *Let S be a set of n points in $\mathbb{E}^d$. For any $\varepsilon > 0$, there exists a structure that uses $O(n)$ storage and $O(n^{1+\varepsilon})$ preprocessing time, such that the number of points in S in a query half space can be counted in $O(n^{1-1/d+\varepsilon})$ time.*

Remark 2.14 The results given in this subsection can be improved slightly. By choosing r differently and doing a more careful analysis, it is possible to obtain $O(n^{1-1/d}(\log n)^{O(1)})$ query time. With some extra tricks, the query time can be reduced even further. However, this does not carry over to the kind of multi-level structures that we will use, so we refer the reader to [80] for further details.

2.6.2   Multi-Level Partition Trees

It is important to note that a partition tree in fact gives us the set of all points that lie in a query half-space as the disjoint union of $O(n^{1-1/d+\varepsilon})$ canonical subsets: When we search with a hyperplane h in a partition tree we visit a number of nodes; at each node we have a simplicial partition, some of whose simplices are contained in h^+. The subset of S containing the points above h is exactly the disjoint union of the canonical subsets corresponding to these simplices. If we are only interested in the number of points in the query half-space it suffices to store the cardinalities of these canonical subsets; when we want to answer a more complicated query we can store these subsets in a secondary structure, leading to multi-level partition trees.

Let us consider the simplex range counting problem, where we want to count the number of points in a query simplex. Observe that a simplex is the intersection of $d+1$ half-spaces. Hence, we can solve the problem with a multi-level partition tree[3], where each of the $d + 1$ levels filters out those points that lie in a particular half-space. The design of this multi-level structure is very similar to the design of multi-level segment trees and multi-level trees based on cuttings. We give a brief description of

---

[3]Actually, partition trees that are based on Theorem 2.12 can handle a simplex range query directly, because the crossing number of a simplex is $(d+1)$ times that of a hyperplane, but we want to illustrate the principle of multi-level partition trees here.

the method. We already know how to count the points in one half-space. To count the number of points in the intersection of two half-spaces, we build a partition tree on the points and we store each canonical subset S_i in an associated structure, which is also partition tree. At the nodes in such a second-level tree we store the cardinalities of the corresponding canonical subsets. To count the number of points in $h_1^+ \cap h_2^+$ we search with h_1 in the main tree. This gives us the points in h_1^+ in $O(n^{1-1/d+\varepsilon})$ canonical subsets S_i. For each subset S_i we query the corresponding second-level tree with h_2, giving us the number of points in $S_i \cap h_2^+$. Adding up these numbers, we obtain the final answer.

The main tree is built using a fine simplicial partition of size $O(r_1)$, and the associated trees with a fine simplicial partition of size $O(r_2)$. For the associated structures we can achieve, for any $\varepsilon' > 0$, a query time of $O(n^{1-1/d+\varepsilon'})$ by choosing r_2 appropriately. Hence, the total query time $Q(n)$ satisfies

$$Q(n) = O(r_1) \times O((n/r_1)^{1-1/d+\varepsilon'}) + O((r_1)^{1-1/d})O(n/r_1),$$

where $\varepsilon' > 0$ can be made arbitrarily small by choosing r_2 appropriately. From the First Recurrence Lemma it follows that for any $\varepsilon > 0$ we can obtain $O(n^{1-1/d+\varepsilon})$ query time, by choosing r_1 sufficiently large. To analyze the amount of storage $M(n)$, we note that the sets S_i in a simplicial partitioning are disjoint. Moreover, the associated structure that stores a set S_i has linear size. Hence, the sum of the sizes of the associated structures over all nodes on a fixed level of the main tree is linear. If we choose the parameter r_1 of the main tree to be a constant, then the depth of the tree is $O(\log n)$, leading to a total amount of storage of $O(n \log n)$. However, by choosing r_1 to be n^δ we can decrease the depth of the tree. More precisely, we set $r_1 = n_0^\delta$ at every node ν in the main tree, where n_0 is the total number of points in S, not the number of points that we consider at node ν. Now the depth becomes constant and the total amount of storage is $M(n) = O(n)$.

This scheme can be extended to counting the number of points in the intersection of three or more half-spaces in a straightforward manner. The query time remains $O(n^{1-1/d+\varepsilon})$ and the amount of storage remains linear. In general, we can use the scheme to select all the points that are above (or below) a number of hyperplanes in a small number of canonical subsets. How these canonical subsets are treated depends on the application.

2.6.3 Trade-offs and Dynamization

In the previous subsection we have seen a method for the problem of selecting points in half-spaces that achieves a good (that is, near to optimal) query time for the case of linear storage. By combining this result with the results of Section 2.5 on cuttings, we can get a trade-off between the query time and the amount of storage. The basic idea is quite simple: we start with the linear storage structure, and we switch to the logarithmic-query time structure when the number of points in a subtree falls below some parameter $\hat{n}$. Clearly, by setting $\hat{n} = n$ we obtain the fast structure based on cuttings, and by setting $\hat{n} = 1$ we obtain the linear size data structure described in the previous subsection. As it turns out, this scheme also matches Chazelle's lower bounds [16] for other values of $\hat{n}$. We confine ourselves to saying that by choosing $\hat{n}$ to be roughly $(m/n)^{1/(d-1)}$ we can obtain, for any $\varepsilon > 0$, a structure that uses $O(m^{1+\varepsilon})$ storage and has $O(n^{1+\varepsilon}/m^{1/d})$ query time. (In fact, the ε can be omitted from the storage if $m = O(n^{d-\delta})$ for some fixed $\delta > 0$.)

Next we turn our attention to the dynamic maintenance of these partition trees. Dynamic partition trees have also been described by Agarwal and Sharir [6]; their dynamic partition trees are based on a slightly different (and slightly more complicated) structure than the one we use. We describe the method only briefly, referring the reader to Overmars [95, Section 7.3] for further details on the technique that we use. A searching problem on a set S is called *decomposable* if for any disjoint subsets $S_1, S_2 \subset S$ with $S_1 \cup S_2 = S$, the answer to a query on the set S can be computed in constant time from the answers to the query on the sets S_1 and S_2. Clearly, the selection of points lying in a query half-space is a decomposable searching problem. Thus we may use the *logarithmic method* to turn our static partition tree into a structure that supports insertions. Since the extra logarithmic factor that the logarithmic method entails are subsumed under the n^{ε}-factor, the query time of the new structure will remain $O(n^{1+\varepsilon}/m^{1/d})$, and the insertion time becomes $O(m^{1+\varepsilon}/n)$. The method can also deal with deletions, provided that we can show how to perform *weak deletions* efficiently. We refer to [95] for a precise definition of weak deletions; for our purposes it suffices to use the fact that deletions that do not increase the future query, deletion and insertion time and the amount of storage are weak. For partition trees it is quite easy to give an algorithm for weak deletions: we just remove the point to be deleted from all the canonical subsets that contain the point. To this end we maintain for every point the canonical subsets that contain it; alternatively, we can search with the point in the partition tree to find the subsets. Notice that in many applications a canonical subset will be stored in an associated structure; in such cases we need, of course, to be able to perform a weak deletion on the associated structure. It remains to bound the number of canonical subsets in which any point in S is contained; this will determine the time that we need for a weak deletion. For linear size partition trees this number is constant; for the structures based on cuttings this number is $O(n^{d-1+\varepsilon})$ if we use the random sampling technique to construct the cuttings. This implies that for the trade-off structure described above that uses $O(m^{1+\varepsilon})$ storage, the number of subsets containing any point in S is $O(m^{1+\varepsilon}/n)$. Here we note that the random sampling method can be made deterministic (see e.g. [78]), which ensures that the update time will be deterministic.

Summarizing the results of this section we obtain the following result.

Theorem 2.15 *Let S be a set of n points in $\mathbb{E}^d$. For any $\varepsilon > 0$, and any $n \leqslant m \leqslant n^d$, there exists a partition tree that uses $O(m^{1+\varepsilon})$ storage, such that it is possible to select the points that lie in a query half-space in $O(n^{1+\varepsilon}/m^{1/d})$ canonical subsets, which can be found in the same amount of time. Insertions and deletions can be performed in $O(m^{1+\varepsilon}/n)$ time.*

Remark 2.16 For the two extreme cases ($m = n$ and $m = n^d$) slightly better results are possible. In particular, one can get rid of the ε at the cost of some extra logarithmic factors. But in order to keep the description simple and uniform, we do not elaborate this point.

2.7 Two Geometric Tools

We close this chapter with the discussion of two geometric transforms, namely the duality transform and the Plücker transform for lines in three-dimensional space. These transforms—which are well known in classical geometry—have proved to be extremely useful in computational geometry.

2.7.1 Duality

Duality transforms, that map points to hyperplanes and vice versa, have been known a long time to geometers. The specific duality transform that we use is described in Edelsbrunner [50, Section 1.4]. Let $p = (p_1, p_2, \ldots, p_d)$ be a point in $\mathbb{E}^d$. Then the dual of p, denoted by p^*, is the non-vertical hyperplane defined by the equation

$$p^* : \quad x_d = 2p_1 x_1 + 2p_2 x_2 + \cdots + 2p_{d-1} x_{d-1} - p_d.$$

Furthermore, the dual $(p^*)^*$ of p^* is defined to be p. Notice that the transform is not defined for vertical hyperplanes. In our applications, this is not a serious problem. We can either treat the vertical hyperplanes separately, or we can use a symbolic perturbation scheme [127] to deal with them. Hence, we ignore the problem in the rest of this book.

The important property of the dual transform is that it is *incidence preserving* and *order preserving*: a point p is contained in a hyperplane h if and only if the point h^* is contained in the hyperplane p^*, and p is above h if and only if h^* is above p^*.

It is worthwhile to mention that the dual transform can be extended to other objects than points and hyperplanes. The dual of a segment s in the plane will be particularly useful to us. The dual s^* of s is a double wedge that is bounded by the duals of the endpoints of s. Notice that these two lines define two double wedges; s^* is the one that does not contain any vertical line. With this duality, a line l intersects a segment s if and only if its dual point l^* lies in the double wedge s^*. Moreover, which of the two wedges of s^* contains l^* gives us some extra information. Consider l to be directed from left to right. Then the wedge containing l^* tells us whether s is hit 'from above' or 'from below' by l. See Figure 2.8, where the lines l_1 and l_2 hit s from different sides when directed from left to right, so their duals are in different wedges of s^*.

2.7.2 Plücker Coordinates

Another geometric transform that we will use is the Plücker transform for lines in space. Plücker coordinates, or Grassmann coordinates, can be defined not only for lines in three-dimensional space, but for any k-dimensional linear variety ($0 \leqslant k \leqslant d - 1$) in d-dimensional space, see e.g. [71]. We will only use them for lines in $\mathbb{E}^3$, however, so we limit our discussion to this case. Stolfi [116] was the first to discuss Plücker coordinates in the context of computational geometry. Their algorithmic possibilities were first exploited in 1989 by Chazelle et al. [22]. Since then, Plücker coordinates have become an important tool for geometric problems in 3-space.

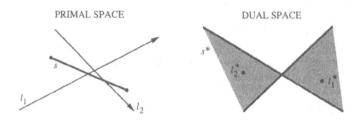

Figure 2.8: A segment and two lines, and their duals.

A line in $\mathbb{E}^3$ can be specified by four parameters. Hence, we can represent a line in $\mathbb{E}^3$ as a point in $\mathbb{E}^4$. However, this representation has some unpleasant characteristics: the set of lines in $\mathbb{E}^3$ that intersect or are parallel to a given line becomes a quadratic hypersurface in $\mathbb{E}^4$, and the position of a query line with respect to a set of given lines is determined by the position of the corresponding point in an arrangement of hypersurfaces. Such arrangements of hypersurfaces are hard to handle explicitly; for example, preprocessing it for fast point location is difficult. Fortunately, there is another representation of lines that does not have these disadvantages, namely by Plücker coordinates and coefficients. Of course, this representation has its cost: the price we pay is an increase of the dimension by one. As we shall see in the sequel, this is almost for free.

Let l be a directed line in 3-space, and let $a = (a_1, a_2, a_3, a_4)$ and $b = (b_1, b_2, b_3, b_4)$ be two different points on l, written in homogeneous coordinates with $a_4, b_4 > 0$, such that l is directed from a to b. Then the *Plücker point* $\pi(l)$ of l is the following point in 6-space

$$\pi(l): \quad (\pi_{12}, \pi_{13}, \pi_{14}, \pi_{23}, \pi_{24}, \pi_{34}),$$

where the Plücker coordinates π_{ij} are defined as

$$\pi_{ij} = a_i b_j - a_j b_i, \quad 1 \leqslant i < j \leqslant 4.$$

Similarly, the *Plücker plane* $\varpi(l)$ of l is given by the equation

$$\varpi(l): \quad \pi_{34}x_1 - \pi_{24}x_2 + \pi_{23}x_3 + \pi_{14}x_4 - \pi_{13}x_5 + \pi_{12}x_6 = 0.$$

Because any positive scalar multiple of $\pi(l)$ is also a Plücker point for l (resulting from different choices of a and b), we can view Plücker coordinates as homogeneous coordinates in projective oriented 5-space. Hence, for our purposes we can treat $\pi(l)$ and $\varpi(l)$ as a point and a hyperplane in 5-space. However, not all points in 5-space are Plücker points of lines in 3-space. A necessary and sufficient condition is that the point lies on the so-called *Plücker hypersurface* or Grassmann manifold of lines in 3-space, which is a quadratic hypersurface Π defined by the equation

$$\Pi: \quad x_1 x_6 - x_2 x_5 + x_3 x_4 = 0.$$

Plücker coordinates have some useful properties. One property that is easily checked is that a line l_1 intersects or is parallel to another line l_2 if and only if $\pi(l_1) \in \varpi(l_2)$. However, it is not true that l_1 lies above l_2 if and only if $\pi(l_1)$ lies on a particular side of $\varpi(l_2)$. But the position of $\pi(l_1)$ relative to $\varpi(l_2)$ corresponds to the *orientation of l_1 relative to l_2*, as follows. Project l_1 and l_2 onto the xy-plane, and suppose that l_1 and l_2 are directed such that $\overline{l_1}$ crosses $\overline{l_2}$ from left to right. (Left and right are defined relative to an observer standing on $\overline{l_2}$ and looking into the direction of the directed line $\overline{l_2}$.) If l_1 passes above l_2 then we can say that l_1 is oriented clockwise with respect to l_2, and if l_1 passes below l_2 then l_1 is oriented counterclockwise with respect to l_2. See Figure 2.9(a), where l_1 is oriented clockwise with respect to l_2. Plücker coordinates in fact express this relation: l_1 is oriented clockwise with respect

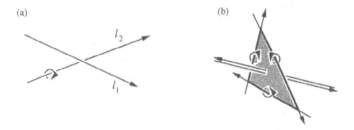

Figure 2.9: Relative orientation.

to l_2 if and only if $\pi(l_1)$ lies on one particular side of $\varpi(l_2)$. We denote this half-space in 5-space containing the Plücker points of all lines that are oriented clockwise with respect to a given line l by $\varpi(l)^+$, and the other half-space by $\varpi(l)^-$.

The above gives us an efficient way to test whether a line intersects a triangle t. If we direct the lines l_1, l_2 and l_3 through the edges of t, then any line that intersects t is oriented in one of two specific ways with respect to these three lines, depending on its own direction. For example, we can direct l_1, l_2 and l_3 such that a line intersects t if and only if it is oriented clockwise with respect to all three lines, or it is oriented counterclockwise with respect to all three lines. See Figure 2.9(b). This means that the Plücker point of a line l that intersects t is either contained in $\varpi(l_1)^+ \cap \varpi(l_2)^+ \cap \varpi(l_3)^+$, or it is contained in $\varpi(l_1)^- \cap \varpi(l_2)^- \cap \varpi(l_3)^-$. Now consider a set of n triangles. The arrangement in 5-space induced by the Plücker planes of the $3n$ directed lines through the edges of the triangles gives us a lot of information about the triangles. Specifically, if we locate the Plücker point corresponding to a query line in this arrangement, then we know exactly which triangles are intersected by the line. Using Theorem 2.7, we can thus answer in $O(\log n)$ time the question how many triangles are intersected by a query line, after $O(n^{5+\varepsilon})$ preprocessing. (It is fairly straightforward to compute the numbers that have to be associated with the leaves in the point location structure.) But we can do even better, if we realize that we only do point location queries with points that lie on the Plücker hypersurface Π. The latter fact implies that we are only interested in the cells of the arrangement that are intersected by Π. The collection of these cells is called the *zone* of Π in the arrangement. It has been shown by Aronov and Sharir [10] (see also [103]) that the total complexity of the zone of Π is $O(n^4 \log n)$. Clearly, when we construct the cutting and we take a random sample R of the hyperplanes of size r, we only have to consider the $O(r^4 \log r)$ simplices of the triangulated arrangement $\mathcal{A}(R)$ that are intersected by Π. Therefore the storage and preprocessing time $T(n)$ of the point location structure satisfies

$$T(n) = O(nr^4 \log r) + O(r^4 \log r)T(n/r),$$

which solves to $T(n) = O(n^{4+\varepsilon})$ by the First Recurrence Lemma. Thus the extra dimension that Plücker coordinates bring along does not appear in the bounds. Note that we have to use random sampling for the preprocessing, because we do not have a bound on the number of simplices in Chazelle's deterministic cuttings [18] that are intersected by the Plücker hypersurface.

Theorem 2.17 *For any $\varepsilon > 0$, it is possible to preprocess a set of n triangles in $\mathbb{E}^3$ in $O(n^{4+\varepsilon})$ randomized time into a data structure that uses $O(n^{4+\varepsilon})$ storage, such that we can select all triangles that are intersected by a query line in $O(\log n)$ time in $O(\log n)$ canonical subsets.*

The $O(r^4 \log r)$ bound on the complexity of the zone of the Plücker hypersurface is not the only thing that makes going to five-dimensional space less expensive than it seems. A second piece of luck comes from the Upper Bound Theorem [50] on the maximal complexity of convex polytopes. Consider the following query problem: Store a set $L = \{l_1, \ldots, l_n\}$ of lines such that we can determine efficiently whether a query line l lies above all lines in L. This problem arises in one of the ray shooting problems that we study in Part B. Chazelle et al. [22] gave an efficient solution based on Plücker coordinates to this problem, which we describe next. Suppose that l and the lines in L are directed such that the projection $\bar{l}$ of l onto the xy-plane crosses all the lines in L from left to right; we say that l is *consistently oriented* with respect to the lines in L. If this is the case, then l is above all the lines if and only if l is oriented clockwise with respect to each of them. The latter condition is equivalent to $\pi(l)$ lying in the convex polytope $\varpi(l_1)^+ \cap \cdots \cap \varpi(l_n)^+$. By the Upper Bound Theorem, the complexity of this polytope is $O(n^{\lfloor d/2 \rfloor}) = O(n^2)$, which implies that the polytope can be preprocessed in $O(n^{2+\varepsilon})$ time for point location queries using random sampling. Again, the extra dimension involved in the use of Plücker coordinates does not appear in the bound, since the maximal complexity of a polytope in 5-space is the same as in 4-space. But of course we cannot direct all the lines L in such a way that any query line is consistently oriented with respect to L. We give a brief description of the solution proposed by Chazelle et al. [22] to this problem. Direct all lines in L from left to right. Let $[-\pi/2, \pi/2)$ be the set of possible slopes of any line. Consider all the lines in L that have greater slope than the query line l. If we direct l from left to right, then l is consistently oriented with respect to exactly these lines. Similarly, l is consistently oriented with respect to the lines that have smaller slope if we direct l from right to left. The idea is to select the lines that have greater slope than l in a constant number of canonical subsets, and to select the lines with smaller slope in a constant number of canonical subsets. For each canonical subset we can use the solution sketched above: compute the Plücker polytope that corresponds to the subset and preprocess it for point location. The selection of the lines with, say, greater slope than a query line in a constant number of canonical subsets is just a one-dimensional half-space range query. In the section on cuttings we have seen how to design and analyze such data structures. Thus the following theorem will come as no surprise.

Theorem 2.18 (Chazelle et al. [22]) *For any $\varepsilon > 0$ it is possible to preprocess a set of n lines in $\mathbb{E}^3$ in $O(n^{2+\varepsilon})$ randomized time into a data structure that uses $O(n^{2+\varepsilon})$ storage, such that we can decide in $O(\log n)$ time whether a query line lies above all the given lines.*

Part B

Ray Shooting

Chapter 3

Introduction

The ray shooting problem is to preprocess a set of objects into a data structure such that the first object that is hit by a query ray can be computed efficiently. As we have seen in Chapter 1, this is an important problem in computer graphics. For this reason the ray shooting problem is one of the more widely studied problems in computational geometry [1, 6, 7, 14, 20, 27, 32, 36, 39, 41, 64]. In the plane this has led to many efficient solutions. For example, ray shooting inside a simple polygon can be answered in $O(\log n)$ time with a structure that uses $O(n)$ storage [20, 27]. For general planar scenes, where the objects are arbitrary line segments, one can obtain $O(\log n)$ query time using $O(n^2\alpha^2(n))$ storage [1], or $O(\sqrt{n}\log n)$ query time using $O(n\log^2 n)$ storage [32]. Here $\alpha(n)$ is the extremely slowly growing functional inverse of Ackermann's function. It is also possible to obtain a trade-off between the query time and the amount of storage [1]. These results are believed to be close to optimal. Even for curved segments efficient data structures have been developed [7].

In three-dimensional space, however, not much was known until recently. When the origin of the query ray is fixed, results have been obtained for polyhedral terrains—$O(\log n)$ query time using $O(n\alpha(n)\log n)$ storage [36]—, for horizontal axis-parallel rectangles—$O(\log n)$ query time using $O(n\log n)$ storage [14]—, and for horizontal triangles—$O(\sqrt{n}\log^2 n)$ query time using $O(n\log^3 n)$ storage [1]. However, these problems can hardly be considered truly three-dimensional. A result that has a more three-dimensional flavor is due to Schmitt, Müller and Leister [111]; they show that a set of axis-parallel polyhedra in space can be preprocessed into a data structure of size $O(n^3\log^{O(1)} n)$ such that a query with an arbitrary ray takes $O(\log^3 n)$ time. They also present an $O(n\log^{O(1)} n)$ size structure with $O(n^{0.695})$ query time. An important step towards a solution to the general problem was taken in 1989 by Chazelle et al. [22]. They showed how to use Plücker coordinates to preprocess a polyhedral terrain into a structure of size $O(n^{2+\epsilon})$ such that ray shooting queries with arbitrary rays take $O(\log^2 n)$ time. The introduction of Plücker coordinates and the development of cuttings for three- and higher dimensional arrangements have opened the way to attack the general ray shooting problem in three-dimensional space. In 1990 Pellegrini [101] was the first to present a structure for ray shooting with arbitrary rays in a set of non-intersecting triangles. His structure uses $O(n^{5+\epsilon})$ preprocessing, and it has $O(\log n)$ query time[1]. After that several new results on three-dimensional ray shooting have been obtained, many of which can be found in this part of the thesis, in particular in Chapter 7.

Let us briefly discuss the ray shooting results that are presented in subsequent chapters.

We start in Chapter 4 by presenting a general strategy for devising efficient data structures for ray shooting problems. Many of the structures to be presented in later chapters are based on this general strategy. The remaining three chapters study the three-dimensional ray shooting problem in various settings of increasing difficulty. Within each chapter, several different classes of objects are studied, including axis-parallel, c-oriented and arbitrary polyhedra. Most of the results from Chapters 5 and 6 are taken from [39, 42, 43]; Chapter 7 is based on [41].

Chapter 5 considers the easiest setting of the problem: ray shooting from a fixed point. The first class of objects that we consider are axis-parallel polyhedra. We present a data structure based on segment trees, which uses $O(n \log n)$ preprocessing time and storage and has $O(\log n)$ query time. Using some more machinery, the structure can be adapted to handle c-oriented polyhedra. The query time then becomes $O(c^2 \log n)$. Arbitrary polyhedra are studied at the end of Chapter 5. Not surprisingly, this class is the most difficult one of the three classes, and the complexity of the data structure is higher than in the previous cases. In particular, we need $O(n^2 \alpha(n))$ storage to obtain $O(\log n)$ query time. The structure that achieves this performance is very simple. The use of multi-level partition trees is required to obtain a structure that uses less storage. For any $\varepsilon > 0$ and any $n \leqslant m \leqslant n^2$, we can obtain a structure that uses $O(m^{1+\varepsilon})$ storage and has $O(n^{1+\varepsilon}/\sqrt{m})$ query time.

In Chapter 6 we allow the starting point of the query ray to vary, but we fix its direction. It turns out that this problem is not much more difficult than the previous one. Indeed, the structures that are developed for ray shooting from a fixed point often can be adapted to this case, with only a small increase in preprocessing and query time. Thus we obtain structures with $O(\log^2 n)$ and $O(c^2 \log^2 n)$ query time for axis-parallel and c-oriented polyhedra, respectively, using $O(n \log^2 n)$ preprocessing. We also study the classes of axis-parallel and c-oriented curtains in this chapter. These classes admit slightly more efficient solutions than axis-parallel or c-oriented polyhedra: after $O(n \log n)$ preprocessing one can obtain $O(\log n)$ query time, or $O(c \log n)$ query time in the c-oriented case. This will prove useful in Part D to obtain a fast algorithm for hidden surface removal for axis-parallel and c-oriented polyhedra. Finally, arbitrary polyhedra are studied. Using similar methods as in the case of shooting from a fixed point we obtain, for any $\varepsilon > 0$ and any $n \leqslant m \leqslant n^3$, a structure with $O(n^{1+\varepsilon}/m^{1/3})$ query time that uses $O(m^{1+\varepsilon})$ storage.

The last chapter of this part deals with arbitrary query rays. First, we study axis-parallel and c-oriented polyhedra. For any fixed $\varepsilon > 0$, a data structure is presented that uses $O(n^{2+\varepsilon})$ storage, and that has $O(\log n)$ resp. $O(c \log n)$ query time. The tools to obtain this result are dualization and cuttings. This result was obtained independently by Pellegrini [102]. Next we turn our attention to arbitrary curtains. By

---

[1]Pellegrini's solution in [101] is incomplete. An improved and corrected result can be found in [102].

using Plücker coordinates and random sampling we obtain a structure with $O(\log n)$ query time that uses $O(n^{2+\varepsilon})$ storage. Besides their usefulness in Parts C and D, curtains are also useful for solving other ray shooting problems. As an example we show how to use curtains for ray shooting in polyhedral terrains, and in horizontal triangles whose angles are greater than some fixed minimum angle. The last class that we treat is the class of arbitrary polyhedra; the structure that we present has $O(\log n)$ query time and uses $O(n^{4+\varepsilon})$ preprocessing time and storage. Agarwal and Sharir [6], Pellegrini [102] and Agarwal and Matousek [4] obtained similar results, using different methods.

For many of the structures discussed above we also give dynamic variants. These structures can be updated efficiently when a polyhedron is inserted into or deleted from the set in which we shoot. Their query time is almost as good as their static counterparts.

Before we proceed, let us introduce some notation that we use throughout this part. A *ray* is a half-line that is directed away from its endpoint. More formally, a ray—which we always denote by ρ—is defined as $\rho = p + \lambda \vec{d}$, $\lambda \geqslant 0$. The point p is called the *starting point* of ρ, and the vector $\vec{d}$ is called the *direction* of ρ. The set of objects in which we shoot is denoted by S. The answer to a ray shooting query in a set S of objects with ray ρ is the first object in S that is intersected by ρ—that is, the object whose intersection with ρ is closest to p—, or *NIL* if ρ misses all objects. The answer to the query is denoted by $\Phi_\rho(S)$. When we are shooting in a set S of polyhedra, we are not satisfied with the first polyhedron that is hit, but we want to know the first face of a polyhedron that is hit. Abusing our notation slightly, we still denote this face by $\Phi_\rho(S)$. We will not worry about whether a face is intersected when the ray intersects the boundary of the face, or about what happens when the query ray and some face lie in a common plane; the way such cases should be handled depends on the application. Instead, we simply assume that such degenerate cases do not occur. For all structures described, the adaptations that are needed are straightforward.

Chapter 4

A General Strategy

In this chapter we describe a general strategy to answer ray shooting queries. Many of the structures to be presented in subsequent chapters are based on this strategy. The basic idea is to reduce a ray shooting query to a logarithmic number of so-called line intersection queries. A more advanced method uses only a constant number of so-called intersection sequence queries and priority intersection queries. However, the latter approach sometimes entails a slight increase in the preprocessing time and the amount of storage.

We also present a simple decomposition scheme such that ray shooting queries in polyhedra can be answered by ray shooting in a set of simpler shapes (rectangles, quadrilaterals and triangles, to be precise). This will make the description of many of the data structures of subsequent chapters easier.

4.1 The Strategy

Let S be the set of objects that we want to preprocess for fast ray shooting queries. The strategy aims at reducing ray shooting queries to *line intersection queries* in certain subsets $S' \subset S$: 'Does a query line intersect at least one object in S'?' The idea of the method is as follows. Suppose an order on the objects exists such that if a ray hits an object o before it hits an object o', then o comes before o' in the ordering. Then $\Phi_\rho(S)$ is the first object in this ordering that is intersected. Thus, if at least one of the objects in the first half of the ordered set is intersected then the answer is in the first half, otherwise it has to be sought in the second half. Let us also assume that an object is intersected by ρ if and only if it is intersected by the line $l(\rho)$ containing ρ, that is, no object lies behind the starting point of the ray. Then we can compute $\Phi_\rho(S)$ by performing a binary search on the ordered set, guided by the answers to certain line intersection queries. To construct a data structure that is based on this strategy, we clearly need an order which is valid for any permissible query ray. Unfortunately, even for rays with a fixed origin such an order does not always exist. A second problem concerns the assumption that an object is intersected by ρ if and only if it is intersected by $l(\rho)$. Obviously, this is not true in general. As we shall see, both problems can be solved with the help of cuttings.

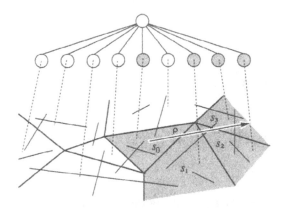

Figure 4.1: A cutting and the corresponding tree. The simplices of $\mathcal{I}_\rho$ are shaded.

Let us make this rough idea more concrete. Let $\Xi(S)$ be a $(1/r)$-cutting for the set S of size $b(r)$, where r does not depend on $|S|$. Thus space is partitioned into $b(r)$ simplices such that each simplex is intersected by at most n/r objects. For a simplex s, let $S_s = \{o \cap s : o \in S\}$, that is, S_s contains the objects that intersect s, restricted to the part that lies inside s. Because we will treat the sets S_s recursively, we will take care that the objects in S_s are in the same 'class' as the objects in S. For example, if S consists of axis-parallel polyhedra then so will S_s. The cutting $\Xi(S)$ imposes an order on the objects, as follows. Any ray ρ starts in some simplex s_0 and then traverses a number of other simplices $s_1, \ldots, s_c$. We call $\mathcal{I}_\rho = s_0, \ldots, s_c$ the *intersection sequence* of ρ. An intersection sequence consists of a *starting simplex* s_0, and a *suffix* $\mathcal{S}_\rho = s_1, \ldots, s_c$. Note that $\mathcal{I}_\rho$ can be determined in $O(b(r) \log b(r))$ time, which is constant because r is a constant. Clearly, if i is the smallest index such that S_{s_i} contains at least one object that is intersected by ρ, then S_{s_i} must contain $\Phi_\rho(S)$. (Observe that we use here the fact that the simplices in the cutting and the query ray are convex objects.) We then continue the search in S_{s_i}. For the simplices of the suffix $\mathcal{S}_\rho$ we can use a line intersection query to test if ρ intersects at least one object, since ρ passes through the whole simplex. For the starting simplex, however, we cannot do this. This means that we always have to search in S_{s_0}. Let us illustrate this with the two-dimensional example of Figure 4.1. In this example, we are ray shooting in a set of arbitrary line segments in the plane. The ray starts in simplex s_0, so we must (recursively) search in S_{s_0}. The ray ρ misses all segments in s_0. Thus we perform a line intersection query with $l(\rho)$ on the set S_{s_1}. Since this query yields *FALSE*, we try S_{s_2}. We are more lucky this time, and so we know that a recursive search in S_{s_2} will give us the answer to the query.

Let us define LINEINTERSECTION(S', l) to be *TRUE* when at least one object in S' is intersected by l and *FALSE* otherwise. For a subset $S' \subset S$ let $\mathcal{D}_I(S')$ be a data structure that answers line intersection queries, that is, $\mathcal{D}_I(S')$ evaluates

LINEINTERSECTION(S', l) when queried with the line l. The above observations lead to the following data structure for ray shooting in the set S.

- If the number of objects in S is some small enough constant then the structure consists of one leaf, where we store all objects in S.

- Otherwise, let $\Xi_{root(\mathcal{T})} = \Xi(S)$ be a $(1/r)$-cutting of size $b(r)$ for S, where r is a constant. The structure is a tree $\mathcal{T}$ of branching degree $b(r)$. There is a one-to-one correspondence between the children of $root(\mathcal{T})$ and the simplices of $\Xi_{root(\mathcal{T})}$.

    - At $root(\mathcal{T})$ we explicitly store the simplices of $\Xi_{root(\mathcal{T})}$.

    - For each simplex $s \in \Xi(S)$ an associated structure $\mathcal{D}_I(S_s)$ is stored for line intersection queries in the set $S_s = \{o \cap s : o \in S\}$.

- The child ν_s of $root(\mathcal{T})$ that corresponds to simplex s is the root of a recursively defined structure on the set $S(\nu_s) = S_s$.

Algorithm 4.1 is a recursive procedure that reports the first object in $S(\nu)$ that is hit by a query ray ρ, or *NIL* if no object is hit. The first time it is called with $\nu = root(\mathcal{T})$. When we recurse in a node that corresponds to a simplex of the suffix, we can replace ρ by the line $l(\rho)$ that contains ρ, and we switch to a subprocedure LINESHOOT.

Algorithm 4.1
Input: A node ν in the main tree $\mathcal{T}$, and a query ray ρ.
Output: $\Phi_\rho(S(\nu))$.
1. **if** ν is a leaf
2.   **then** Compute $o = \Phi_\rho(S(\nu))$ by checking all objects in $S(\nu)$. **return** o.
3.   **else** Determine the intersection sequence $\mathcal{I}_\rho = s_0, \ldots, s_c$ of ρ in Ξ_ν.
4.   Compute $o = \Phi_\rho(S(\nu_{s_0}))$ recursively.
5.   **if** $o \neq NIL$
6.   **then return** o
7.   **else** $i \leftarrow 1$
8.   **while** $i \leqslant c$ **and** LINEINTERSECTION$(S_{s_i}, l(\rho))$=*FALSE*
9.   **do** $i \leftarrow i + 1$
10.   **if** $i = c + 1$
11.   **then return** *NIL*
12.   **else return** LINESHOOT(ν_{s_i}, ρ)

The subprocedure LINESHOOT is defined as follows.

procedure LINESHOOT(ν, ρ)
1. **if** ν is a leaf
2.   **then** Compute $o = \Phi_\rho(S(\nu))$ brute-force. **return** o.
3.   **else** Determine the intersection sequence $\mathcal{I}_\rho = s_0, \ldots, s_c$ of ρ in Ξ_ν.
4. $i \leftarrow 0$
5.   **while** $i \leqslant c$ and LINEINTERSECTION($S_{s_i}, l(\rho)$)=$FALSE$
6.   **do** $i \leftarrow i + 1$
7.   **if** $i = c + 1$
8.   **then return** *NIL*
9.   **else return** LINESHOOT(ν_{s_i}, ρ)
end LINESHOOT

Let us summarize the ingredients that we need to use this approach for ray shooting in a set S of n objects. Suppose we can

(i) construct a $(1/r)$-cutting of size $b(r)$, where $r \geqslant 2$ is constant, in time $T_C(n, r)$,

(ii) preprocess any subset $S' \subset S$ of size m in time $T_I(m)$ into a data structure $\mathcal{D}_I(S')$ that uses $M_I(m)$ storage and answers line intersection queries in time $Q_I(m)$.

If these two conditions are fulfilled then we have the following general theorem.

Theorem 4.2 *Let S be a set of n objects satisfying the conditions given above. Then we can answer ray shooting queries in S in time $Q(n) = O(Q_I(n) \log n)$ with a structure whose preprocessing time $T(n)$ satisfies*

$$T(n) = T_C(n, r) + b(r)T_I(n/r) + bT(n/r)$$

and whose storage requirement $M(n)$ satisfies

$$M(n) = O(b(r)) + b(r)M_I(n/r) + b(r)M(n/r).$$

Proof: The correctness of the method follows from the discussion above, so let us prove the bounds on the preprocessing and query time. The recurrences for the preprocessing time and the amount of storage immediately follow from the definition of the data structure. To prove the query time, we define $Q_R(n)$ to be the time that Algorithm 4.1 takes when at least one object is intersected, and $Q'_R(n)$ to be the time taken when no object is intersected. Clearly $Q(n) = \max(Q_R(n), Q'_R(n))$. Finally, we define $Q_L(n)$ to be the time taken by the procedure LINESHOOT. From the query algorithm we infer the following recurrences:

$$Q_R(n) \leqslant \max \begin{cases} 1 + Q_R(n/r) \\ Q'_R(n/r) + b(r)Q_I(n/r) + Q_L(n/r) \end{cases}$$
$$Q'_R(n) \leqslant Q'_R(n/r) + b(r)Q_I(n/r)$$
$$Q_L(n) \leqslant b(r)Q_I(n/r) + Q_L(n/r)$$

Because $r \geqslant 2$ is a constant, the recurrences for $Q'_R(n)$ and $Q_L(n)$ solve to $O(Q_I(n) \log n)$. Hence, $Q_R(n) = O(Q_I(n) \log n)$ as well. $\qquad\square$

Remark 4.3 Notice that the fact that the ordering is obtained by taking a $(1/r)$-cutting is not important. The only thing we need is that S can be partitioned into (not necessarily disjoint) subsets $S_1, \ldots, S_b$ that can be ordered with respect to a given query ray, and that we can treat ρ as a full line with respect to all but one of these subsets. To improve the intuition for the strategy and to avoid introducing too much notation we have chosen not to describe the method in its full generality. Moreover, all structures in this thesis that are based upon this strategy use cuttings to impose the order, except the structure of Section 6.2.3.

4.2 Reducing the Query Time

When we take a closer look at the proof of Theorem 4.2 we see that the query time is in fact $O(Q_I(n)b(r) \log n / \log r)$, which is $O(Q_I(n) \log n)$ if r is a constant. Hence, we might be able to reduce the query time by choosing r larger, for example, $r = n^\varepsilon$ for some $\varepsilon > 0$. Choosing this value of r will make the factor $O(\log n / \log r)$ constant. But this cannot be done just like that, since we also have a factor $b(r)$ in the query time (in our applications $b(r) = O(r^k)$ for some small constant k). Hence, we would like to remove the dependency on $b(r)$ in the query time.

Let $S_\rho = s_1, \ldots, s_c$ be the suffix of a query ray ρ. We need to determine the smallest index i such that S_{s_i} contains at least one object that is intersected by $l(\rho)$. In the previous section b was a constant, so we could perform line intersection queries in S_{s_1}, S_{s_2}, and so on, until the line intersection query yields *TRUE* for the first time. When b is not a constant we cannot afford this brute approach. What we need is a data structure for so-called *priority intersection queries*: 'Given an ordered collection of subsets of S, report the first subset that contains at least one object that is intersected by a query line l.' The query time of this structure $\mathcal{D}_{PI}(S)$ should be independent of the number of subsets. Notice that we need such a structure for every possible suffix S that can occur.

Another problem is the determination of the intersection sequence $\mathcal{I}_\rho$. When b is constant a brute-force method is fine, but now we need to be more careful. So we need a data structure for *intersection sequence queries*: 'Compute the intersection sequence $\mathcal{I}_\rho$ of a query ray ρ.' More precisely, the answer to an intersection sequence query consists of two pointers: a pointer to the child ν_{s_0} corresponding to the starting simplex of ρ, and a pointer to the data structure $\mathcal{D}_{PI}(S_\rho)$ for priority intersection queries in the suffix of ρ.

Putting all this together we obtain the following structure.

- If the number of objects in S is some small enough constant then the structure consists of one leaf, where we store all objects in S.

- Otherwise, let $\Xi_{root(\mathcal{T})} = \Xi(S)$ be a $(1/r)$-cutting of size $b(r)$ for S, where $r = n^\varepsilon$ for some $\varepsilon > 0$. The structure is a tree $\mathcal{T}$ of branching degree $b(r)$. There is a one-to-one correspondence between the children of $root(\mathcal{T})$ and the simplices of $\Xi_{root(\mathcal{T})}$.

  - At $root(\mathcal{T})$ an associated structure $\mathcal{D}_{IS}$ is stored that can answer intersection sequence queries for the cutting $\Xi_{root(\mathcal{T})}$.

- For every possible suffix $S = s_1, \ldots, s_c$ an associated structure $\mathcal{D}_{PI}(\mathcal{S})$ is stored that can answer priority intersection queries on the ordered collection of sets $S_{s_1}, \ldots, S_{s_c}$.

- The child ν_s of *root*$(\mathcal{T})$ that corresponds to simplex s is the root of a recursively defined structure on the set $S(\nu_s) = S_s$.

The query algorithm is the same as in the previous section, except for two things: in line 3 of Algorithm 4.1 and of procedure LINESHOOT we determine the intersection sequence using the structure $\mathcal{D}_{IS}$, and lines 8–9 of Algorithm 4.1 and lines 5–6 of LINESHOOT—the **while**-loops with line intersection queries—are replaced by one priority intersection query. The conditions under which we can reduce the query in the way sketched above are thus as follows. Suppose we can

(i) construct a $(1/r)$-cutting of size $b(r)$, where $r = n^\varepsilon$ and $\varepsilon > 0$, in time $T_C(n, r)$,

(ii) preprocess any ordered collection of k subsets of S with total size m in time $T_{PI}(k, m)$ into a data structure $\mathcal{D}_{PI}(S')$ that uses $M_{PI}(k, m)$ storage and answers priority intersection queries in time $Q_{PI}(m)$,

(iii) preprocess $\Xi(S)$ in time $T_{IS}(b(r))$ into a data structure $\mathcal{D}_{IS}$ of size $M_{IS}(b(r))$ that answers intersection sequence queries in time $Q_{IS}(b(r))$.

If all this is possible, then we have the following theorem.

Theorem 4.4 *Let S be a set of n objects satisfying the conditions given above. Then we can answer ray shooting queries in time $O(\log n + Q_{PI}(n) + Q_{IS}(n))$ with a structure whose preprocessing time $T(n)$ satisfies*

$$T(n) = T_C(n, r) + T_{IS}(b(r)) + n_S T_{PI}(b(r), n) + b(r) T(n/r)$$

and whose storage requirement $M(n)$ satisfies

$$M(n) = M_{IS}(b(r)) + n_S M_{PI}(b(r), n) + b(r) M(n/r),$$

where $r = n^\varepsilon$ for some $\varepsilon > 0$ and n_S is the total number of different suffixes in $\Xi(S)$.

Proof: The recurrences for $T(n)$ and $M(n)$ follow directly from the definition of the data structure. The bound on the query time can be proved in the same way as in Theorem 4.2. The only difference is that, because $r = n^\varepsilon$, we do not get a multiplicative log-factor for $Q_{PI}(n)$ and $Q_{IS}(n)$, provided that $Q_{PI}(n)$ and $Q_{IS}(n)$ are at least logarithmic. See the Second Recurrence Lemma. If $Q_{PI}(n)$ and $Q_{IS}(n)$ are sublogarithmic then we can bound the query time by $O(\log n)$.   □

Some readers may worry about the factor n_S in the bounds on the preprocessing. But the number of different suffixes is not as large as it might seem. Indeed, in our applications it will be polynomial in $b(r)$, which is in turn polynomial in r. Since $r = n^\varepsilon$ and we can choose $\varepsilon > 0$ arbitrarily small, this means that the increase in preprocessing is of the form $n^{\varepsilon'}$, for an $\varepsilon' > 0$ that we can make arbitrarily small by choosing ε sufficiently small.

Remark 4.5 The strategy presented in this chapter is not the only general strategy to tackle ray shooting problems. Indeed, Agarwal and Matoušek [3] have described a general strategy based on *parametric search*. Their method reduces ray shooting queries to *segment emptiness queries*: 'Given a query segment, does it intersect any of a set of objects?' Thus our first method leads to a simpler subproblem, namely one with lines instead of segments. Moreover, the trick that we used to reduce the query time does not seem to work in their method. Finally, our method method has a geometrical nature which makes it more intuitive than the parametric search method. Our method does have one disadvantage, however: it is not clear how to trade query time for storage. Such a trade-off, which is important several contexts, is possible if one uses the method of Agarwal and Matoušek.

4.3 Decomposing Polyhedra

It is somewhat easier to describe ray shooting structures for a set of triangles than it is to describe structures for polyhedra. Hence, we present a decomposition scheme that reduces the latter problem to the first one. Similar decomposition schemes are given for axis-parallel and c-oriented polyhedra.

4.3.1 Axis-Parallel Polyhedra

Let S be a set of axis-parallel polyhedra with n vertices in total. Each face of a polyhedron in S is a rectilinear polygon. It can be decomposed into axis-parallel rectangles in a standard manner, as follows. Consider a face f that is parallel to the xy-plane. We compute the vertex-edge pairs of f that are visible in the y-direction, that is, the pairs consisting of a vertex and an edge that can be connected by a segment that is parallel to the y-axis and whose interior does not intersect the boundary of f. The segments connecting these pairs decompose f into axis-parallel rectangles. See Figure 4.2. Faces that parallel to the xz-plane or to the yz-plane are decomposed in a similar manner.

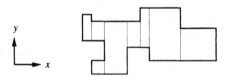

Figure 4.2: Decomposing an axis-parallel face.

The resulting set of rectangles can be partitioned into three subsets: a subset S_1 of rectangles that are parallel to the xy-plane, a subset S_2 of rectangles that are parallel to the xz-plane and a subset S_3 of rectangles that are parallel to the yz-plane. Trivially, $\Phi_\rho(S)$ can be computed in constant time from $\Phi_\rho(S_1)$, $\Phi_\rho(S_2)$ and $\Phi_\rho(S_3)$. This leads to the following lemma.

Lemma 4.6 *Ray shooting queries in a set of axis-parallel polyhedra with n vertices in total can be answered in asymptotically the same bounds as ray shooting queries in a set of n axis-parallel rectangles that are parallel to the xy-plane.*

Proof: The structure for computing $\Phi_\rho(S)$ consists of three separate structures, one for each subset S_i. Note that if we can build a ray shooting structure for axis-parallel rectangles that are parallel to the xy-plane, then we can also build a structure for axis-parallel rectangles that are parallel to, for example, the xz-plane, using a suitable transformation.

The decomposition described above results in $O(n)$ rectangles and it can be computed in linear time by the recent algorithm of Chazelle [17]. (In fact, in all our applications any good old $O(n \log n)$ time algorithm will do as well.) Let $M'(m)$ denote the storage used by the ray shooting structure on a set of m rectangles. Then the storage used by the structure for axis-parallel polyhedra is $M(n) = M'(|S_1|) + M'(|S_2|) + M'(|S_3|)$. Because $\sum_{1 \leqslant i \leqslant 3} |S_i| = O(n)$, and $M'(m)$ can be assumed to be at least linear, we have $M(n) = O(M'(n))$. By the same argument the amount of preprocessing does not increase. Because the query time is non-decreasing in n and $\Phi_\rho(S)$ can be computed in constant time from $\Phi_\rho(S_1)$, $\Phi_\rho(S_2)$ and $\Phi_\rho(S_3)$, the query time remains the same. $\qquad\square$

4.3.2   *c*-Oriented Polyhedra

A decomposition scheme similar to that for axis-parallel polyhedra can be used to decompose the faces of c-oriented polyhedra. Consider a face in a c-oriented set S of polyhedra. If the face is parallel to the yz-plane, then we decompose it by connecting the vertex-edge pairs that are visible in the y-direction with a segment that is parallel to the y-axis. For faces that are not parallel to the yz-plane we do the same thing, using segments that are parallel to the yz-plane. Observe that because the face is not parallel to the yz-plane, the orientation of the segments is uniquely determined. This decomposition results in a set of quadrilaterals, each of which has two edges that are parallel to the yz-plane, called the *left edge* and the *right edge*, and two other edges, called the *top edge* and the *bottom edge*. Some of the quadrilaterals will be degenerate; their left or right edge has length zero and thus they are triangles.

After the decomposition, we partition the set of quadrilaterals into a number of subsets according to the orientation of their top edges and bottom edges: two quadrilaterals are in the same subset if and only if their top edges are parallel and their bottom edges are parallel. Notice that if the orientation of the top edges in a subset is different from the orientation of the bottom edges, then all faces in the subset are parallel to each other and all left and right edges are also parallel to each other. If the orientation of the top edges is equal to the orientation of the bottom edges, however, this is not the case. Therefore we partition these subsets further according to the inclination of the faces: two faces are put into the same subset if and only if they are parallel to each other. We are now left with $2c(c-1)$ subsets $S_1, S_2, \ldots, S_{2c(c-1)}$: we have $c(c-1)$ subsets where the orientation of the top edge is different from the orientation of the bottom edges, and each of the c subsets where the orientation is equal is partitioned further into $c - 1$ new subsets.

Lemma 4.7 Let $\mathcal{D}$ be a data structure for ray shooting in a set of n quadrilaterals, whose top edges are parallel to each other, whose bottom edges are parallel to each other, and whose left and right edges are parallel to each other, which has $Q(n)$ query time. Then a data structure exists for ray shooting in a set of c-oriented polyhedra with n vertices in total, which has $O(c^2 Q(n/c^2))$ query time, and whose preprocessing time and storage requirement are asymptotically the same as the preprocessing time and storage requirement of $\mathcal{D}$.

Proof: The bound on the storage follows in the same way as in Lemma 4.6. The bound on the query time also follows, if we realize that $Q(n)$ can be assumed to be sublinear.   □

4.3.3   Arbitrary Polyhedra

As in the axis-parallel and c-oriented case, it is also convenient to decompose the faces of an arbitrary polyhedron into manageable pieces. Because we cannot hope for any nice properties of these pieces, our decomposition is just a triangulation of the faces. Hence, the decomposition can be carried out in linear time [17]. The following lemma is straightforward.

Lemma 4.8 Ray shooting queries in a set of arbitrary polyhedra with n vertices in total can be answered in asymptotically the same bounds as ray shooting queries in a set of n triangles.

Chapter 5

Ray Shooting from a Fixed Point

In this section we study the ray shooting problem for rays emanating from a fixed starting point p. This is sometimes referred to as the *implicit hidden surface removal problem* [1, 64]. An interesting application of this problem are so-called *mouse-click locations*, where one selects an object on the screen with a mouse. Conceptually, this means that one shoots a ray from a point above the scene into a direction that is perpendicular to the screen. The first object that is hit is the object that is selected. Since parallel rays that start above the scene can be viewed as rays that start at a fixed point at infinity, the data structures for the ray shooting problem that we present in this section are object-space solutions to the mouse-click location problem.

5.1 Axis-Parallel Polyhedra

As mentioned above, structures for ray shooting from a fixed point can be used for mouse-click locations. In some cases the objects to be selected are windows, which can be regarded as axis-parallel rectangles that are parallel to the screen. In this section we study a slightly more general problem, where the objects are axis-parallel polyhedra. We present a static structure that has $O(\log n)$ query time and that uses $O(n \log n)$ storage. We also give a dynamic structure with $O(\log^2 n)$ query time and $O(\log^2 \log \log n)$ update time, using $O(n \log^2 n)$ storage. Similar results, using a different method, have been obtained by Bern [14].

5.1.1 A Static Structure

As a warm-up exercise, we start with the case of ray shooting from a fixed point in a set of horizontal segments in the plane. After that we show how to reduce the three-dimensional problem to a number of two-dimensional problems. Plugging in the solution to our warm-up exercise then gives us the desired structure.

The planar case

Suppose we are given a set S of n line segments in the yz-plane that are parallel to the y-axis, which we want to preprocess for efficient ray shooting from a fixed point p lying in the yz-plane as well. Furthermore, let p have greater z-coordinate than all the segments. (This is not at all necessary, but it fits in better with the three-dimensional case to be discussed later.) Thus we only have to consider rays that are directed downward. To simplify the description, we assume that $p_z > 0$.

Project the $2n$ endpoints of the segments onto the y-axis, with p as the center of projection. The projected endpoints partition the y-axis into (at most) $2n + 1$ intervals. Let y_ρ be the intersection of the y-axis with a query ray ρ starting at p. The interval that contains y_ρ uniquely determines which segments are hit by ρ and, hence, it also determines $\Phi_\rho(S)$. Let us merge adjacent intervals that correspond to the same answer into one, and label each interval with its corresponding segment. What we obtain is a representation of the so-called *upper envelope of S with respect to p*, denoted by $\mathcal{E}(S, p)$. See Figure 5.1 for an illustration. (Normally, upper envelopes

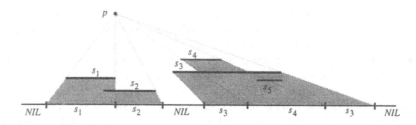

Figure 5.1: The upper envelope of a set of segments with respect to point p.

are defined using an orthogonal projection, and the additive 'with respect to point p' is omitted. In other words, the point that is the center of projection is usually assumed to be located at infinity.) We can determine $\Phi_\rho(S)$ by computing the intersection point y_ρ of ρ with the y-axis and searching with this point in the representation of $\mathcal{E}(S, p)$.

The upper envelope of a set of horizontal segments in the plane can be computed in optimal $O(n \log n)$ time. However, if we have sorted lists of the z-coordinates of the endpoints of the segments and of the y-coordinates of the projected endpoints, then the upper envelope can be computed in linear time, as has been shown by Asano et al. [11]. Since this fact will be useful in the next section we summarize the above as follows.

Lemma 5.1 *Let S be a set of n segments in the yz-plane that are parallel to the y-axis and let p be a fixed point with greater z-coordinate than the segments in S. Ray shooting queries in S from point p can be answered by searching with the intersection point of ρ with the y-axis in $\mathcal{E}(S, p)$. If we have sorted lists of the z-coordinates of the endpoints in S and the y-coordinates of the projected endpoints, then $\mathcal{E}(S, p)$ can be computed in linear time.*

The three-dimensional case

Next we turn our attention to the three-dimensional case. The idea is to use a segment tree to reduce the dimension of the problem by one, so that we can use the planar structure described above.

By Lemma 4.6 we can restrict our attention to ray shooting in a set S of n axis-parallel rectangles that are parallel to the xy-plane. To simplify the discussion, let us split S into a set S^+ containing all faces whose z-coordinate is greater than p_z, and a set S^- containing all faces whose z-coordinate is smaller than p_z. (The assumption that the ray is not in a common plane with any face implies that we do not have to consider faces that are at the same height as p.) We will describe the structure that stores S^-. S^+ is stored in a similar structure. Notice that we always have to query in only one of the two structures: if ρ is directed downward then we only query in S^-, and if ρ is directed upward then we only query in S^+.

Project the rectangles in S^- onto the xy-plane, with p as the center of projection. (We assume without loss of generality that p lies above the xy-plane.) The projection $\overline{f}$ of a face $f \in S^-$ can be written as $[\overline{f}_x : \overline{f}'_x] \times [\overline{f}_y : \overline{f}'_y]$. We call $[\overline{f}_x : \overline{f}'_x]$ the x-*interval of* $\overline{f}$, and we call $[\overline{f}_y : \overline{f}'_y]$ the y-*interval of* $\overline{f}$. Observe that a ray ρ starting at p intersects a face f if and only if the intersection point (x_ρ, y_ρ) of ρ with the xy-plane is contained in $\overline{f}$. We build a segment tree $\mathcal{T}$ for the x-intervals of the rectangles $\overline{f}$. This tree serves to filter out those rectangles whose x-interval contains x_ρ. For a node ν in $\mathcal{T}$, let $S^-(\nu)$ be its canonical subset. Consider a node ν on the search path of x_ρ in $\mathcal{T}$. Because we already know that $x_\rho \in [\overline{f}_x : \overline{f}'_x]$ for faces $f \in S^-(\nu)$, the problem that remains to be solved at ν has become two-dimensional: we can project the whole scene orthogonally onto the yz-plane and use Lemma 5.1 to solve the planar problem. Thus we store at node ν the upper envelope of the set $\{[\overline{f}_y : \overline{f}'_y] \times f_z : f \in S^-(\nu)\}$ with respect to (p_y, p_z). This upper envelope is stored as an ordered list $\mathcal{L}_\nu$ of y-intervals; each interval is labeled with the face $f \in S^-(\nu)$ whose z-coordinate is maximal at that interval. To compute $\Phi_\rho(S^-(\nu))$ we have to search with y_ρ in the list $\mathcal{L}_\nu$. To facilitate this search we apply fractional cascading to the tree $\mathcal{T}$ with the associated lists $\mathcal{L}_\nu$.

Let us summarize the above with a succinct description of the data structure for the set S^- and of the query algorithm.

- The rectangles of S^- are stored in a segment tree $\mathcal{T}$ on the x-intervals of their projections onto the xy-plane.

  - At each node ν in $\mathcal{T}$, the upper envelope of $\{[\overline{f}_y : \overline{f}'_y] \times f_z : f \in S^-(\nu)\}$ with respect to (p_y, p_z) is stored in a sorted list $\mathcal{L}_\nu$.

- Fractional cascading is applied to the tree $\mathcal{T}$ with the associated lists $\mathcal{L}_\nu$.

This structure is queried with rays that are directed downward, according to the following algorithm.

Algorithm 5.2
Input: The query ray ρ.
Output: $\Phi_\rho(S^-)$.
1. Compute the intersection point (x_ρ, y_ρ) of ρ with the xy-plane.
2. Search with x_ρ in $\mathcal{T}$.
3. **for** each node ν on the search path
4.   **do** Search with y_ρ in $\mathcal{L}_\nu$ to compute $f_\nu = \Phi_\rho(S^-(\nu))$.
5. Compute $f = \Phi_\rho(S^-)$ by taking the nearest of all faces f_ν found in line 4.
6. **return** f.

Recall that we have a similar structure, with a similar query algorithm, that allows us to compute $\Phi_\rho(S^+)$ for rays that are directed upward.

The preprocessing is done as follows. Consider the structure for the set S^-. First, we build the skeleton of the segment tree $\mathcal{T}$ for the set of x-intervals of the rectangles in S^-. Then we sort the rectangles in order of increasing z-coordinate, and we insert the x-intervals with their corresponding rectangles in this order into $\mathcal{T}$. This way we obtain at each node ν a list of the rectangles in $S^-(\nu)$ that is sorted on z-coordinate. In the same way we can obtain a sorted list of the y-coordinates of the intervals $[\overline{f}_y : \overline{f}'_y]$. Using these lists, we construct at each node ν the upper envelope of $\{[\overline{f}_y : \overline{f}'_y] \times f_z : f \in S^-(\nu)\}$ with respect to (p_y, p_z) in linear time, using the algorithm of Asano et al. [11]. Finally, we set up the fractional cascading structure for $\mathcal{T}$ and the lists $\mathcal{L}_\nu$.

We conclude with the following theorem.

Theorem 5.3 *Ray shooting queries from a fixed point in a set of axis-parallel polyhedra with n vertices in total can be answered in $O(\log n)$ time with a structure that uses $O(n \log n)$ storage. The structure can be built in $O(n \log n)$ time.*

Proof: The theorem follows if we can prove that the structure for the set S^- uses the $O(|S^-| \log |S^-|)$ storage and preprocessing time, and that the query algorithm correctly reports $\Phi_\rho(S^-)$ in $O(\log |S^-|)$ time. The construction of the skeleton of $\mathcal{T}$, the presorting and the insertion of the the x-intervals take $O(|S^-| \log |S^-|)$ time. The lists $\mathcal{L}_\nu$ can be computed in total time $\sum_{\nu \in \mathcal{T}} O(|S^-(\nu)|)$, by Lemma 5.1. Because in a segment tree $\sum_{\nu \in \mathcal{T}} |S^-(\nu)| = O(|S^-| \log |S^-|)$, and the application of fractional cascading does not increase the preprocessing time asymptotically, the storage and preprocessing time are as claimed.

Now consider a query in this structure. From the way the x-intervals are defined and from the properties of segment trees it follows that only those faces can be hit by ρ that are stored at nodes ν on the search path of x_ρ in $\mathcal{T}$. By definition of the upper envelope, a search in $\mathcal{L}_\nu$ gives us the face with largest z-coordinate of all faces in $S^-(\nu)$ such that $y_\rho \in [\overline{f}_y : \overline{f}'_y]$. Because we already know that $x_\rho \in [\overline{f}_x : \overline{f}'_x]$, and because p is above all faces in S^- this must be $\Phi_\rho(S^-(\nu))$. This proves the correctness of the query algorithm. During the query algorithm we search in each of the $O(\log |S^-|)$ lists $\mathcal{L}_\nu$ that are on the search path to x_ρ. Fractional cascading allows us to do each search in constant time, after spending $O(\log |S^-|)$ time for the first search. This proves the query time. $\square$

5.1.2   Dynamization

The structure that we just described cannot be updated efficiently. The problem lies
in the associated structures that we store at each node in the segment tree: the upper
envelope of a set of segments can change drastically after an update. Indeed, both
insertions and deletions can cause a linear structural change in the envelope. Hence,
we must use a different associated structure.

The planar case

We return to the planar case, where we are given a set S of line segments in the
yz-plane that are parallel to the y-axis, and a fixed starting point p that has greater
z-coordinate than the segments. Let y_ρ be the intersection point of ρ with the y-axis.
Note that ρ intersects a segment if and only if y_ρ is contained in the projection of the
segment onto the y-axis, when p is taken to be the center of projection. We build
a segment tree $\mathcal{T}$ on these projected segments. For a node ν in $\mathcal{T}$, let $S(\nu)$ be its
canonical subset. Now consider a node ν on the search path of y_ρ in $\mathcal{T}$. The first
segment in $S(\nu)$ that is hit is the segment with largest z-coordinate. In a static setting
it would be sufficient to store this segment, but when the aim is a dynamic structure
we must store all segments in $S(\nu)$. We store these segments in a list $\mathcal{L}_\nu$ that is sorted
on the z-coordinate. A query with ray ρ is performed by searching with y_ρ in $\mathcal{T}$ and
selecting the segment with largest z-coordinate at all nodes ν on the search path. Of
the $O(\log n)$ segments that we find, we select the one that is hit first.

Lemma 5.4 *Let S be a set of n segments in the yz-plane that are parallel to the
y-axis. Ray shooting queries in S from a fixed point p can be answered in $O(\log n)$
time with a structure that uses $O(n \log n)$ storage. A segment can be inserted into or
deleted from S in $O(\log n \log \log n)$ amortized time.*

Proof: This immediately follows from Theorem 2.4 on dynamic fractional cascading
in segment trees. The fact that the query time is $O(\log n)$ instead of $O(\log n \log \log n)$
is a consequence of the fact that we always want the segment with largest z-coordinate
at each node ν on the search path. $\square$

The three-dimensional case

The dynamic solution to the two-dimensional ray shooting problem can be turned into
a three-dimensional structure in the same way as in the static case. For the reader's
convenience we give a brief description of the resulting structure.

By Lemma 4.6 we can limit ourselves to ray shooting in a set S of n axis-parallel
rectangles that are parallel to the xy-plane. Consider the subset $S^- \subset S$ of rectangles
whose z-coordinate is smaller than p_z. The rectangles in S^- are stored in a segment

tree $\mathcal{T}$ on the x-intervals of their projection onto the xy-plane. The two-dimensional problem that we have at each node ν in $\mathcal{T}$ is solved using Lemma 5.4.

A query is performed as follows. We search with x_ρ in $\mathcal{T}$, and for each node ν on the search path we perform a two-dimensional query in $\mathcal{T}_\nu$. Of the $O(\log n)$ faces that we find, we select the one closest to p.

To insert a polyhedron $\mathcal{P}$ (of constant complexity), we decompose the faces of $\mathcal{P}$ into rectangles and we insert each rectangle into the right structure. The insertion of a rectangle into tree $\mathcal{T}$ is done in the normal way: locate the nodes ν such that the x-interval of the rectangle contains I_ν but not $I_{parent(\nu)}$, and insert the rectangle into $\mathcal{T}_\nu$. Of course, we rebalance the structure if necessary. Deletions are done analogously. See Sections 2.3 and 2.4 for more details on updates on (multi-level) segment trees.

This leads to the following theorem.

Theorem 5.5 *Ray shooting queries from a fixed point in a set of axis-parallel polyhedra with n vertices in total can be answered in $O(\log^2 n)$ time with a structure that uses $O(n \log^2 n)$ storage. An axis-parallel polyhedron of constant complexity can be inserted into or deleted from the structure in $O(\log^2 n \log \log n)$ amortized time.*

Proof: The correctness of the reduction to the two-dimensional problem follows in the same way as in the static case. The theorem now follows from Lemma 5.4 and standard arguments for multi-level segment trees and dynamic fractional cascading. See Theorems 2.2 and 2.4.   □

5.2 c-Oriented Polyhedra

In this section we study ray shooting from a fixed point p in a set of c-oriented polyhedra. Recall that a c-oriented set of polyhedra is a set of polyhedra whose edges have only c different orientations. It turns out that, with some extra machinery, the solution of the previous section can be extended so that it also handles c-oriented polyhedra, leading to a query time of $O(c^2 \log n)$.

5.2.1   A Static Structure

Let S be a set of n quadrilaterals such that all top edges are parallel to each other, all bottom edges are parallel to each other, and all left and right edges are parallel to each other. If we can solve the ray shooting problem for S then we can also solve it for a set of c-oriented polyhedra, by Lemma 4.7. To simplify the notation, let us assume that the left and right edges are parallel to the y-axis and that the bottom edges are parallel to the x-axis. Notice that this implies that the quadrilaterals are parallel to the xy-plane. As in the axis-parallel case, we split S into a subset S^+ of faces whose z-coordinate is greater than p_z, and a subset S^- of faces whose z-coordinate is smaller than p_z. These subsets are stored in two separate structures of which we only have to query one for a given query ray ρ. Let us describe the structure that stores the set S^-, which is queried with downward directed rays.

Project the faces in S^- onto the xy-plane with p as the center of projection. We denote the projection of a quadrilateral f by $\overline{f}$. Because the left and right edge of f are parallel to the y-axis, so are the left and right edges of $\overline{f}$. We define $\overline{f}_x$ and $\overline{f}'_x$ to be the x-coordinates of the projected left and right edge, and we call $[\overline{f}_x : \overline{f}'_x]$ the x-interval of $\overline{f}$. We store the faces in a segment tree $\mathcal{T}$ according to their x-intervals. For a node ν in $\mathcal{T}$, let $S^-(\nu)$ be its canonical subset. Recall from Section 2.3 that for the projection $\overline{f}$ of a face $f \in S^-(\nu)$ we restrict our attention to its relevant part, which is the part that lies inside the slab corresponding to ν. Note that the relevant part is again a quadrilateral. We decompose the projection $\overline{f}$ of each face $f \in S^-(\nu)$ into two pieces, a rectangular piece $r_\nu(\overline{f})$ and a triangular piece $t_\nu(\overline{f})$, as follows. Let

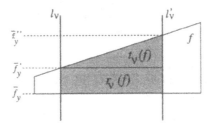

Figure 5.2: Decomposing (the relevant part of) a quadrilateral.

$I_\nu = [x_\nu : x'_\nu] \times [-\infty : \infty]$ be the slab corresponding to ν, and let l_ν be the vertical line $x = x_\nu$ that bounds that slab from the left. Similarly, we define l'_ν to be the line $x = x'_\nu$ that bounds the slab from the right. Finally, define $\overline{f}_y$ to be the y-coordinate of the the projected bottom edge of f—recall that this edge is parallel to the x-axis— and define $\overline{f}'_y$ and $\overline{f}''_y$ to be the y-coordinates of the intersection of the top edge of f with the l_ν respectively l'_ν. See Figure 5.2. We assume without loss of generality that $\overline{f}'_y \leqslant \overline{f}''_y$. The rectangular piece $r_\nu(\overline{f})$ of $\overline{f}$ inside the slab corresponding to ν is $[x_\nu : x'_\nu] \times [\overline{f}_y : \overline{f}'_y]$; the remaining piece of $\overline{f}$ inside the slab is the triangular piece $t_\nu(\overline{f})$. We denote the part of f projecting onto $r_\nu(\overline{f})$ by $r_\nu(f)$, and the part projecting onto $t_\nu(\overline{f})$ by $t_\nu(f)$. Observe that in some cases the part of $\overline{f}$ in the slab already is a triangle, and no splitting needs to be performed.

At each node ν we have two associated structures, one for the set $S^-_R(\nu)$ of rectangular pieces and one for the set $S^-_T(\nu)$ of triangular pieces. The rectangular pieces can be handled as in the axis-parallel case (see Section 5.1): the projection of each piece has a well defined y-interval $[\overline{f}_y : \overline{f}'_y]$, so we can compute the upper envelope of the segments $[\overline{f}_y : \overline{f}'_y] \times f_z$, and store it in a list $\mathcal{L}^R_\nu$.

We now turn our attention to the triangular pieces. Recall that the top edges of the triangles as well as the bottom edges are parallel, and that they exactly span the slab corresponding to ν. In other words, the triangles $t_\nu(f) \in S^-_T(\nu)$ are translates of each other. For a query ray ρ, let the triangle $t_\nu(\rho)$ be defined as follows. Let (x_ρ, y_ρ) be the intersection of ρ with the xy-plane. Rotate any triangle $180°$ around its left vertex and let $t_\nu(\rho)$ be the translate of this rotated triangle that has (x_ρ, y_ρ) as its right vertex. More formally, $t_\nu(\rho)$ is the Minkowski difference of (x_ρ, y_ρ) and a copy of the translates with its left vertex placed at the origin. Now observe that ρ intersects a triangle if and only if the left vertex of that triangle is contained in $t_\nu(\rho)$. See Figure 5.3. Moreover, since all the left vertices lie on the line l_ν, it is even true that

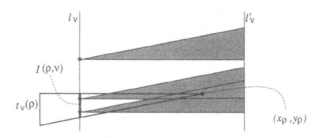

Figure 5.3: $t_\nu(\rho)$ contains the left vertices of the triangles that are intersected by ρ.

ρ intersects a triangle $t_\nu(\overline{f})$ if and only if the left vertex $(x_\nu, \overline{f}'_y)$ of $t_\nu(\overline{f})$ is contained in the intersection $I(\rho, \nu)$ of $t_\nu(\rho)$ with l_ν. Recall that the faces in S^- are parallel to the xy-plane, and that p has greater z-coordinate than all faces. This implies that $\Phi_\rho(S_T^-(\nu))$ corresponds to the triangular piece with largest z-coordinate of all pieces with $(x_\nu, \overline{f}'_y) \in I(\rho, \nu)$. Therefore we define $V_T^-(\nu) = \{(\overline{f}'_y, f_z) : t_\nu(f) \in S_T^-(\nu)\}$, and we store $V_T^-(\nu)$ in a list $\mathcal{L}_\nu^T$ that is sorted on the $\overline{f}'_y$-values. On top of this list we build a structure that enables us to find the point $(\overline{f}'_y, f_z)$ with largest f_z-value that lies between two query elements of $\mathcal{L}_\nu^T$. Gabow, Bentley and Tarjan [60] have shown that a structure exists that answers such a query in constant time, if the positions of the query elements in the list are known.

To perform a query at node ν, we compute the interval $I(\rho, \nu)$ in constant time, and we search with the y-coordinates of this interval in $\mathcal{L}_\nu^T$. We find two elements in $\mathcal{L}_\nu^T$, and we compute the point with largest f_z-value that lies in between these two elements. To speed up the search with the y-coordinates of $I(\rho, \nu)$, we apply fractional cascading to the associated lists $\mathcal{L}_\nu^T$ of the tree $\mathcal{T}$. However, there is one small problem with the application of fractional cascading. The y-coordinate of the upper endpoint of an interval $I(\rho, \nu)$ is always equal to y_ρ, but the y-coordinate of the lower endpoint is not the same at every node ν. Fortunately, this is not a serious problem, as can be seen as follows. Consider a different coordinate system, where the y-axis is replaced with an axis perpendicular to the top edges of the quadrilaterals in S^-. In this coordinate system, the new coordinate of the lower endpoint is the same at every node ν. Hence, we need two fractional cascading structures, one for the upper endpoints where the cascading is done in y-direction, and one for the lower endpoints where the cascading is done in the direction perpendicular to the top edges.

In short, the structure for the set S^- looks as follows.

- The quadrilaterals of S^- are stored in a segment tree $\mathcal{T}$ on the x-intervals of their projections onto the xy-plane.

  - At each node ν in $\mathcal{T}$ we have a list $\mathcal{L}_\nu^R$, storing the upper envelope of the set $\{[\overline{f}_y : \overline{f}_y'] \times f_z : r_\nu(f) \in S_R^-(\nu)\}$ with respect to (p_y, p_z).

  - At each node ν in $\mathcal{T}$, we have a list $\mathcal{L}_\nu^T$ storing the set $V_T^-(\nu) = \{(\overline{f}_y', f_z) : t_\nu(f) \in S_T^-(\nu)\}$. The list is sorted on the $\overline{f}_y'$ values. On top of this list, we have the structure of [60] that allows us to find the vertex with highest f_z-value between two query elements in the list.

- There is a fractional cascading structure in y-direction for the associated lists $\mathcal{L}_\nu^R$, and there are two fractional cascading structures for the lists $\mathcal{L}_\nu^T$, one in y-direction and one in the direction perpendicular to the top edges of the quadrilaterals in S^-.

A query with a downward directed ray ρ is performed according to the following algorithm.

Algorithm 5.6
Input: The query ray ρ.
Output: $\Phi_\rho(S^-)$.
1. Compute the intersection point (x_ρ, y_ρ) of ρ with the xy-plane.
2. Search with x_ρ in $\mathcal{T}$.
3. **for** each node ν on the search path
4.   **do** Search with y_ρ in $\mathcal{L}_\nu^R$ to compute $f_{\nu,R} = \Phi_\rho(S_R^-(\nu))$.
5.   Compute the interval $I(\rho, \nu)$ and locate the endpoints of $I(\rho, \nu)$ in $\mathcal{L}_\nu^T$.
6.   Search in the structure built on top of $\mathcal{L}_\nu^T$ to compute $f_{\nu,T} = \Phi_\rho(S_T^-(\nu))$.
7. Compute f_R, the nearest of the faces $f_{\nu,R}$ found in line 4.
8. Compute f_T, the nearest of the faces $f_{\nu,T}$ found in line 6.
9. **return** $\Phi_\rho(\{f_R, f_T\})$.

The construction of the structure is done in the same way as in the axis-parallel case. To build the structure for S^- we construct the skeleton of the segment tree $\mathcal{T}$. Then we insert the quadrilaterals of S^- into $\mathcal{T}$ and we build the associated structures. To be able to construct the upper envelope of the segments that correspond to the rectangular pieces in $S_R^-(\nu)$ efficiently, we need at each node ν a sorted list of the y-coordinates of the endpoints of these segments, that is, of the values $\overline{f}_y$, $\overline{f}_y'$ and f_z. The values $\overline{f}_y$ and f_z are the same for a quadrilateral f at every node ν where it is stored. Hence, by presorting these edges before they are inserted into $\mathcal{T}$ we can obtain a sorted list of these coordinates at each node ν. To obtain a sorted list of the values $\overline{f}_y'$ at node ν, we note that this value is the y-coordinate of the intersection of the top edge of f with the line l_ν. This implies that the order of these values corresponds to the order of the top edges. Thus the sorted list of the $\overline{f}_y'$ values can be obtained by presorting the top edges. A simple merge of the sorted lists of $\overline{f}_y$ and $\overline{f}_y'$ values results in the sorted list of all y-coordinates that is needed for the construction of the upper envelopes. To construct the associated structures that store the triangular pieces in linear time, we need at each node ν a sorted list of the $\overline{f}'_y$ values; we already have seen how to obtain this list.

Theorem 5.7 *Ray shooting queries from a fixed point in a set of c-oriented polyhedra with n vertices in total can be answered in $O(c^2 \log n)$ time with a structure that uses $O(n \log n)$ storage. The structure can be built in $O(n \log n)$ time.*

Proof: The preprocessing for the structure of Gabow et al. [60] is linear, if we have a sorted list of the points available. In the same way as in the axis-parallel case it then follows that ray shooting queries in a set of quadrilaterals of a fixed type can be answered in $O(\log n)$ time after $O(n \log n)$ preprocessing. The bounds of the theorem follow, because a ray shooting query in a set of c-oriented polyhedra can be answered by answering $O(c^2)$ ray shooting queries in sets of quadrilaterals of a fixed type. See Lemma 4.7.   □

Remark 5.8 In the next section we present a structure for ray shooting into a fixed direction in a set of non-intersecting c-oriented polyhedra. Shooting into a fixed direction is more general than shooting from a fixed point. Hence, we can also use the structure of the next section for the problem studied here, provided that the polyhedra are non-intersecting. The query time of this structure, which is $O(c \log^2 n)$, is superior to the query time of the present structure when $c = \Omega(\log n)$.

5.2.2   Dynamization

To dynamize the structure presented above, we need associated structures that are dynamic. The rectangular pieces can be stored in the same way as in the dynamic axis-parallel case (Section 5.1.2), so let us concentrate on the triangular pieces. Consider the tree $\mathcal{T}$ storing the set S^- of faces whose z-coordinate is smaller than p_z. Recall that the associated structure for the triangular pieces at a node ν in $\mathcal{T}$ stores a set $V_T^-(\nu)$ of points in a two-dimensional space; the structure is used to report the point with largest second coordinate of all points with their first coordinate in a certain query range. A dynamic solution to this problem is the *priority search tree* of McCreight [82]. Priority search trees use a linear amount of storage, and have logarithmic query and update time. This immediately leads to the following theorem.

Theorem 5.9 *Ray shooting queries from a fixed point in a set of c-oriented polyhedra with n vertices in total can be answered in $O(c^2 \log^2 n)$ time with a structure that uses $O(n \log^2 n)$ storage. A c-oriented polyhedron of constant complexity can be inserted into or deleted from the structure in $O(\log^2 n \log \log n)$ amortized time.*

Proof: Observe that the associated structure that stores a set $S_T^-(\nu)$ of triangular pieces uses linear storage and has $O(\log |S_T^-(\nu)|)$ query and update time, whereas the structure that stores the set $S_R^-(\nu)$ of rectangular pieces uses $O(|S_R^-(\nu)| \log |S_T^-(\nu)|)$ storage, has $O(\log |S_R^-(\nu)|)$ query time and $O(\log |S_R^-(\nu)| \log \log |S_R^-(\nu)|)$ update time. Hence, the storage and time bounds are dominated by the associated structures for the rectangular pieces, and we obtain the same bounds as in the dynamic rectangular case.   □

5.3 Arbitrary Polyhedra

We have seen that ray shooting from a fixed point p in a set of axis-parallel or c-oriented polyhedra can be done in logarithmic time using a roughly linear amount of storage. If we are shooting in a set S of arbitrary polyhedra we cannot obtain structures with the same efficiency. Indeed, to obtain $O(\log n)$ query time we will need a quadratic amount of storage, and with near-linear storage we can only achieve roughly $O(\sqrt{n})$ query time. We start with the solution that uses a quadratic amount of storage, since it is simple. Then we take a more sophisticated approach, which leads to a solution that uses a nearly linear amount of storage. This solution also has the possibility to trade storage for query time. Similar results have been obtained by Agarwal [1].

5.3.1 A Simple Structure with Logarithmic Query Time

Let S be a set of polyhedra with n vertices in total. Assume that the starting point p of the query rays has greater z-coordinate than (any point on) the polyhedra in S; if this it not the case we can always partition the polyhedra in S into a part lying above p and a part lying below p, and treat these parts separately. We also assume without loss of generality that p lies above the xy-plane. Obviously, if p is fixed then the answer to a query is uniquely determined by the intersection point (x_ρ, y_ρ) of ρ with the xy-plane. The *locus approach* now suggests to partition the xy-plane into regions where the answer to a query is fixed. A query can then be answered by locating the point (x_ρ, y_ρ) in the resulting subdivision. This subdivision, where each region is labeled with the face that is hit in the region, is called the *visibility map* of S with respect to p, since it corresponds to the view of S when looking from point p. This subdivision is denoted by $\mathcal{M}_p(S)$. Part D is devoted to the topic of computing visibility maps. Here it suffices to say that the complexity of the visibility map is $O(n^2)$ if the polyhedra are non-intersecting, and $O(n^2\alpha(n))$ if the polyhedra are allowed to intersect [53, 88]. Furthermore, $\mathcal{M}_p(S)$ can be computed in $O(n^2)$ time in the non-intersecting case [83], and in $O(n^2\alpha(n))$ time in the intersecting case [51]. After $\mathcal{M}_p(S)$ has been computed it remains to construct a point location structure for it, which can be done in time that is linear in the complexity of the map [17, 73]. All we have to do to compute $\Phi_\rho(S)$ is to locate, in $O(\log n)$ time, the point (x_ρ, y_ρ) in $\mathcal{M}_p(S)$. We obtain the following theorem.

Theorem 5.10 *Ray shooting queries from a fixed point in a set of arbitrary polyhedra with n vertices in total can be answered in $O(\log n)$ time with a structure that uses $O(n^2\alpha(n))$ storage, which can be built in $O(n^2\alpha(n))$ time. If the polyhedra are non-intersecting, then the amount of storage and the preprocessing time reduce to $O(n^2)$.*

Remark 5.11 Observe that the complexity of $\mathcal{M}_p(S)$ need not be quadratic at all. Indeed, it can be only constant. In that case the storage requirement of the data structure decreases accordingly. Moreover, by using the output-sensitive algorithms to be presented in Part D to compute $\mathcal{M}_p(S)$, it is possible to reduce the preprocessing time if the complexity of the map is not too large.

5.3.2   Trade-Offs and Dynamization

If we do not want to spend a quadratic amount of storage, we cannot afford to store the whole visibility map of the scene. Therefore we take the following approach. First, we select the faces that intersect $l(\rho)$, the line containing the query ray ρ, in a small number of canonical subsets. Because we know that $l(\rho)$ intersects every face in such a group, we can treat the faces as planes and the problem reduces to the problem of ray shooting in a convex polyhedron. The latter problem can be solved very efficiently, using an algorithm of Dobkin and Kirkpatrick [48]. Next, we give a more detailed description of the data structure.

For simplicity, let us consider a set S of n (possibly intersecting) triangles. By Lemma 4.8 this is sufficient to solve the general problem. The selection in a small number of canonical subsets of the triangles that are intersected by $l(\rho)$ is done using partition trees (see Section 2.6). To this end we project S onto the xy-plane, with p as the center of projection. Consider the projection $\bar{t}$ of a triangle t. Observe that $\bar{t}$ is exactly the intersection of three half-planes that are bounded by the lines $l_1(\bar{t})$, $l_2(\bar{t})$ and $l_3(\bar{t})$ through the edges of $\bar{t}$. Since lines through p intersect t if and only if they intersect $\bar{t}$, we can use this fact to select the triangles that are intersected by $l(\rho)$. Let us denote the half-planes whose intersection is $\bar{t}$ by $l_1(\bar{t})^+$, $l_2(\bar{t})^+$ and $l_3(\bar{t})^+$. Furthermore, let $L_i(\overline{S}) = \{l_i(\bar{t}) : t \in S\}$, $i = 1, 2, 3$. The triangles that are intersected by $l(\rho)$ can be selected by a three-level partition tree. (In Section 2.6 we described partition trees to select points in a half-plane. By duality, they can also be used to select lines above or below a query point.) The main tree is a partition tree $\mathcal{T}$ on the set $L_1(\overline{S})$. The second-level partition trees $\mathcal{T}_\nu$, which store the canonical subsets $S(\nu)$ of nodes ν in $\mathcal{T}$, are partition trees for the sets $\{l_2(\bar{t}) : t \in S(\nu)\}$. The third-level partition trees $\mathcal{T}_\mu$, which store the canonical subsets $S(\mu)$ of nodes μ in $\mathcal{T}_\nu$, are partition trees for the sets $\{l_3(\bar{t}) : t \in S(\mu)\}$. To select the triangles in S that are intersected by $l(\rho)$ we proceed as follows. Let (x_ρ, y_ρ) be the intersection of $l(\rho)$ with the xy-plane. First, we select all triangles t such that $(x_\rho, y_\rho) \in l_1(\bar{t})^+$. Using the main tree $\mathcal{T}$, we can select these triangles in a number of canonical subsets $S(\nu)$. By searching in the second-level trees $\mathcal{T}_\nu$, we select the triangles with $(x_\rho, y_\rho) \in l_2(\bar{t})^+$. For the triangles t in the canonical subsets $S(\mu)$ that we find in the second-level trees, we know that $(x_\rho, y_\rho) \in l_1(\bar{t})^+ \cap l_2(\bar{t})^+$. The third-level partition trees $\mathcal{T}_\mu$ enable us to filter out the triangles such that $(x_\rho, y_\rho) \in l_3(\bar{t})^+$. Thus the disjoint union of the canonical subsets $S(\xi)$ that we find when we are querying the third-level trees corresponds exactly to the subset of triangles in S that are intersected by $l(\rho)$. We want to store $S(\xi)$ such that we can compute $\Phi_\rho(S(\xi))$ efficiently. Let $H_{S(\xi)}$ be the set of planes containing the triangles in $S(\xi)$, and let $\mathcal{P}(H_{S(\xi)})$ be the convex polyhedron which is the intersection of the half-spaces that contain p and are bounded by planes in $H_{S(\xi)}$. Note that ρ intersects a triangle t if and only if $l(\rho)$ intersects t and ρ intersects the plane through t. Because $l(\rho)$ intersects all the triangles in $S(\xi)$ when we select this canonical subset, $\Phi_\rho(S(\xi))$ is uniquely determined by $\Phi_\rho(H_{S(\xi)})$. Thus it suffices to preprocess $\mathcal{P}(H_{S(\xi)})$ for efficient ray shooting. We could do this using the technique of Dobkin and Kirkpatrick [48], who obtain logaritmic query time with linear storage. However, it is unknown how to dynamize their structure. Fortunately, Agarwal and Matoušek [3] have described another structure for ray shooting in a convex polytope, and their structure is dynamic; it uses $O(n^{1+\varepsilon})$ storage, it has $O(\log^5 n)$ query time and it has $O(n^\varepsilon)$ amortized update time.

Summarizing, we obtain a four-level structure.

- There is a three-level partition tree $\mathcal{T}$. The first level is on the set $L_1(\overline{S})$, the second level uses the set $L_2(\overline{S})$, and the third level uses the set $L_3(\overline{S})$.

    - For each canonical subset $S(\xi)$ of each third level tree we store the convex polyhedron $\mathcal{P}(H_{S(\xi)})$, preprocessed for ray shooting using the structure of Agarwal and Matoušek [3].

The query algorithm is as follows.

Algorithm 5.12
Input: The query ray ρ.
Output: $\Phi_\rho(S)$.
1. Compute the intersection point (x_ρ, y_ρ) of $l(\rho)$ with the xy-plane.
2. Select all triangles t such that $(x_\rho, y_\rho) \in \bar{t}$ using the partition tree $\mathcal{T}$.
3. **for** every canonical subset $S(\xi)$ that is selected
4.   **do** Compute $f_\xi = \Phi_\rho(S(\xi))$ by computing $\Phi_\rho(\mathcal{P}(H_{S(\xi)}))$.
5. Compute $f = \Phi_\rho(S)$ by taking the nearest of the triangles f_ξ found in line 4.
6. **return** f.

We obtain the following theorem.

Theorem 5.13 *For any $\varepsilon > 0$ and any $n \leqslant m \leqslant n^2$, there exists a structure for ray shooting from a fixed point in a set of arbitrary polyhedra with n vertices in total that has $O(n^{1+\varepsilon}/\sqrt{m})$ query time and uses $O(m^{1+\varepsilon})$ storage. A polyhedron of constant complexity can be inserted into or deleted from the structure in $O(m^{1+\varepsilon}/n)$ amortized time.*

Proof: The performance of multi-level partition trees is essentially determined by the least efficient level, which is in our case a two-dimensional partition tree. The bounds follow. See Section 2.6 for a more detailed discussion of the analysis of multi-level partition trees. □

Chapter 6

Ray Shooting into a Fixed Direction

In the previous chapter we studied the ray shooting problem for rays emanating from a fixed point p. In this chapter we allow the ray to start at an arbitrary point, but we fix its direction $\vec{d}$. Observe that rays emanating from a fixed point become parallel when we map the starting point to infinity. Hence, the problem of the previous chapter—where the rays are not only parallel but also start at infinity—is a special case of the problem studied here.

6.1 Axis-Parallel Polyhedra

In this section we present structures for ray shooting into a fixed direction in a set of axis-parallel polyhedra. The first structure that we describe is static and has $O(\log n (\log \log n)^2)$ query time. It uses $O(n \log n)$ storage. Using the general strategy of Section 4.2 we then show how the query time can be reduced to $O(\log n)$ at the cost of an extra $O(n^\varepsilon)$ factor in storage. Finally, we present a dynamic structure with $O(\log^2 n \log \log n)$ query and update time that uses $O(n \log^2 n)$ storage.

6.1.1 A Static Structure

The planar case

Again, we start by studying the planar case: We are given a set S of n segments in the yz-plane that are parallel to the y-axis, and we want to answer ray shooting queries with rays that lie in the yz-plane and have a fixed direction. Since the direction of ρ is fixed—let us assume that the direction is the negative z-direction—the answer to a query is uniquely determined by the starting point p of ρ. Thus we can partition the yz-plane into regions of equal answer; to answer a ray shooting query we only have to locate the region that contains the starting point of ρ.

What is the region where the answer to a shooting query into a fixed direction is some segment $s \in S$? This is exactly the part of the plane that can be reached when we shoot from any point on s into the opposite direction, in our case the positive z-direction. Thus the desired subdivision can be obtained in the following way:

connect each endpoint of a segment in S to the segment that is hit when we shoot
from v into positive z-direction. When no segment is hit, we add a vertical half-line
to the subdivision. What we get is a subdivision consisting of $3n$ axis-parallel line
segments. We refine this into a *rectangular subdivision* (a subdivision of the plane into
axis-parallel rectangles) by shooting from each vertex into the negative z-direction as
well. The result is the *vertical adjacency map* [76] of the set S, denoted by $\mathcal{V}(S)$.
See Figure 6.1 for an illustration. This map has linear size and it can be constructed

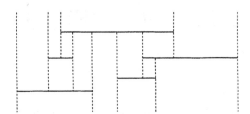

Figure 6.1: A vertical adjacency map.

in $O(n \log n)$ time by a standard sweep line algorithm. To be able to answer queries
quickly we preprocess $\mathcal{V}(S)$ for fast point location. Using a standard point location
structure we can obtain $O(\log n)$ query time, but because this will prove useful when
we study the three-dimensional case, we apply a special point location scheme for
normalized rectangular subdivisions. Thus we normalize the map so that the coordinates of the rectangles are integers in the range $[1 : 2n]$. It will be convenient to
generalize this, and to assume that the coordinates are integers in the range $[1 : U]$. If
the coordinates of the query point are normalized as well, then we can answer a query
in $O((\log \log U)^2)$ time, using the point location method of [45]. This structure uses
a linear amount of storage and can be built in $O(n \log n)$ randomized time, leading to
the following lemma.

Lemma 6.1 *Let S be a set of n segments in the yz-plane that are parallel to the
y-axis, and whose coordinates are integers in the range $[1 : U]$. Ray shooting queries
in S into the negative z-direction can be answered in $O((\log \log U)^2)$ time, with a
structure that uses $O(n)$ storage, provided that the coordinates of the starting point
of the ray are normalized. The structure can be built in $O(n \log n)$ randomized time.*

The three-dimensional case

The solution for the planar case can be used to solve the three-dimensional case, like
in the previous chapter. Recall from Lemma 4.6 that we can restrict our attention
to a set S of n axis-parallel rectangles that are parallel to the xy-plane. To simplify
the notation we assume that the fixed direction of the query rays is the negative
z-direction.

The first level of the structure is identical to the first level of the structure for
ray shooting from a fixed point: it is a segment tree $\mathcal{T}$ on the x-intervals of the

projections of the rectangles onto the xy-plane. The projection is a parallel projection into the fixed direction of the query rays, in our case the z-direction. Let $S(\nu)$ be the canonical subset of node ν in $\mathcal{T}$. At each node ν we are left with a planar problem for the set $\{[f_y : f'_y] \times f_z : f \in S(\nu)\}$, which contains the orthogonal projections of the rectangles in $S(\nu)$ onto the yz-plane. This planar problem can be solved using the techniques that we described above. Recall that the planar point location method requires subdivisions whose coordinates are integers in a fixed range $[1 : U]$. Although we have a point location structure at every node in the segment tree, we normalize the problem only once. Thus the coordinates of the vertical adjacency map $\mathcal{V}_\nu$ stored at node ν in the tree are integers in the range $[1 : 2n]$, not in the range $[1 : 2|S(\nu)|]$. We profit from this, because now the normalization needs to be done only once, whereas it will speed up the time for point location at each node on the search path in $\mathcal{T}$. Note that the size U of the universe does not appear in the bounds on the storage or the preprocessing, so our scheme does not affect these bounds for the total structure.

Summarizing, we have the following structure.

- The rectangles of S are stored in a segment tree $\mathcal{T}$ on the x-intervals of their projections onto the xy-plane.

  - At each node ν in $\mathcal{T}$, the normalized vertical adjacency map $\mathcal{V}_\nu$ of the set of the segments $\{[f_y : f'_y] \times f_z : f \in S(\nu)\}$ is stored, preprocessed for fast point location according to [45].

- The sets of y- and z-coordinates of the projections of the vertices of the rectangles in S are stored in balanced binary search trees $\mathcal{T}_y$ and $\mathcal{T}_z$.

The query algorithm is as follows:

Algorithm 6.2
Input: The query ray ρ.
Output: $\Phi_\rho(S)$.
1. Compute the normalized y-coordinate $\tilde{p}_y$ of p by searching in $\mathcal{T}_y$.
2. Compute the normalized z-coordinate $\tilde{p}_z$ of p by searching in $\mathcal{T}_z$.
3. Search with p_x in $\mathcal{T}$.
4. **for** each node ν on the search path
5.   **do** Compute $f_\nu = \Phi_\rho(S_\nu)$ by locating $(\tilde{p}_y, \tilde{p}_z)$ in $\mathcal{V}_\nu$.
6. Compute $f = \Phi_\rho(S)$ by taking the nearest of the faces f_ν found in line 5.
7. **return** f.

We thus arrive at the following theorem. The proof is the same as the proof of Theorem 5.3, except that we use Lemma 6.1 instead of Lemma 5.1.

Theorem 6.3 *Ray shooting queries into a fixed direction in set of axis-parallel polyhedra with n vertices in total can be answered in $O(\log n(\log \log n)^2))$ time with a structure that uses $O(n \log n)$ storage. The structure can be built in $O(n \log^2 n)$ randomized time.*

6.1.2   Reducing the Query Time

Although it is usually not considered to be a good trade-off, it is possible to shave off the $O((\log \log n)^2)$ factor in the query time at the cost of an extra $O(n^\varepsilon)$ factor in storage. However, in some applications—in particular one of the hidden surface removal algorithms to be presented in Part D—the number of queries can warrant this approach. The reduction in query time is based upon the general strategy presented in Section 4.2.

Let S be a set of n rectangles that are parallel to the xy-plane, and assume that the fixed direction of the query rays is the negative z-direction. To apply Theorem 4.4 we need three ingredients: (i) a $(1/r)$-cutting of (hopefully small) size $b(r)$, (ii) a structure for intersection sequence queries, and (iii) a structure for priority intersection queries. Because the rectangles in S are parallel to the xy-plane it is easy to find a good cutting.

Lemma 6.4 *It is possible to construct a $(1/r)$-cutting $\Xi(S)$ of size $O(r)$ for a set S of n rectangles that are parallel to the xy-plane, in $O(n \log n)$ time.*

Proof: Sort the rectangles according to their z-coordinate and take a plane through every $\lfloor n/r \rfloor$th rectangle. This results in $O(r)$ parallel planes. The *slab* in between two consecutive planes (and the region below the first plane, and the region above the last plane) is a (degenerate) simplex that contains no more than n/r rectangles. (Strictly speaking, if two or more rectangles have the same z-coordinate this is not always correct. Since two rectangles with the same z-coordinate are never intersected by the same ray—and can therefore be ordered arbitrarily—this does not impose any serious difficulties.)   □

An intersection sequence query, which asks for the sequence of simplices of the cutting traversed by ρ, is easily answered for this cutting. This is not only true when the direction of ρ is fixed, but also for general rays. The latter fact will be useful in Section 7.1.

Lemma 6.5 *Intersection sequence queries for the cutting $\Xi(S)$ of Lemma 6.4 can be answered in $Q_{IS}(r) = O(\log r)$ time with a structure that uses $M_{IS}(r) = O(r)$ storage and can be built in time $T_{IS}(r) = O(r \log r)$. Furthermore, n_S, the number of possible suffixes in $\Xi(S)$, is $O(r)$.*

Proof: The structure is a binary search tree on the z-coordinates of the planes of the cutting. Clearly, the amount of storage is $O(r)$ and the preprocessing time is $O(r \log r)$. A search in this tree with the z-coordinate of the starting point of ρ finds the starting slab. If ρ is directed upward then the suffix of ρ consists of the slabs above the starting slab, if ρ is parallel to the xy-plane then the suffix of ρ is empty, and if ρ is directed downward (as it always is in this section) then the suffix of ρ consists of the slabs below the starting slab. Thus there are $O(r)$ possible suffixes, and the suffix of a query ray can be found in $O(\log r)$ time.   □

We are left with the problem of finding a structure for priority intersection queries: given an ordered collection of sets of rectangles that are parallel to the xy-plane, determine the first set that contains a rectangle that is intersected by a query line l. Observe that in our case the ordering of the sets is actually an ordering according to the z-coordinates of the rectangles. Thus the problem can be solved by shooting a ray downward from infinity along l: the first set that contains an intersected rectangle is, of course, the one that contains the first rectangle that is hit. We already noted that ray shooting from infinity into a fixed direction is equivalent to ray shooting from a fixed point. Hence, priority intersection queries can be answered in $Q_{PI}(n) = O(\log n)$ time after $O(n \log n)$ preprocessing, by Theorem 5.3.

This leads to the following theorem.

Theorem 6.6 *For any $\varepsilon > 0$, there exists a structure for ray shooting into a fixed direction in a set of axis-parallel polyhedra with n vertices in total that has $O(\log n)$ a query time and uses $O(n^{1+\varepsilon})$ storage and preprocessing time.*

Proof: According to Theorem 4.4, the query time of the ray shooting structure is $O(\log n + Q_{PI}(n) + Q_{IS}(n)) = O(\log n)$. The preprocessing time $T(n)$ of the structure satisfies

$$T(n) = O(n \log n) + n_S O(n \log n) + O(r \log r) + O(r) T(n/r),$$

where $n_S = O(r)$ is the number of possible suffixes, and $r = n^{\varepsilon'}$. Using the First Recurrence Lemma we see that we can obtain, for any $\varepsilon > 0$, $O(n^{1+\varepsilon})$ preprocessing time, by choosing ε' sufficiently small. Obviously, this is also a bound on the storage requirement. Thus ray shooting queries in a set of n axis-parallel rectangles that are parallel to the xy-plane can be answered within the claimed bounds. By Lemma 4.6 this can be extended to axis-parallel polyhedra. $\square$

Remark 6.7 Notice that the structure for priority intersection queries is more powerful than needed: it tells us not only which is the first set containing an intersected rectangle, but is tells us exactly which rectangle in this set is hit first. Therefore there is no need to search recursively in this set and the query algorithm can be simplified accordingly. This simplification does not influence the asymptotic behavior of the structure, however.

Remark 6.8 The reduction in query time can also be achieved by a somewhat simpler structure. This structure is the same as the structure of Section 6.1.1, except that a segment tree of branching degree n^{ε} is used. An interval in such a segment tree can be stored with at most n^{ε} nodes at each level, leading to a bound on the amount of storage of $O(n^{1+\varepsilon})$. We have chosen to describe the structure that is based on Section 4.2 for two reasons. First, we will need some of the results that are obtained in this subsection later on. Second, the structure we have described is a simple example that demonstrates the applicability of the general technique.

6.1.3 Dynamization

The dynamic structure that we present in this subsection is based on the structure of Section 6.1.1. Recall that this structure consists of a segment tree $\mathcal{T}$, with at each node ν in $\mathcal{T}$ a structure for two-dimensional ray shooting. The two-dimensional problem was solved by locating the starting point of the ray in a rectangular subdivision, using the fast point location method of [45]. The point location scheme of [45] allows updates. Unfortunately, the rectangular subdivisions that we store are vertical adjacency maps, and these maps can change drastically when a segment is inserted or deleted. Hence, a different method is needed to solve the planar problem. But we need not look very far for this new method; the problem can be solved in exactly the same way as the planar problem of Section 5.1.2 was solved. Thus, instead of building the vertical adjacency maps $\mathcal{V}_\nu$ we store the rectangles in $S(\nu)$ in a segment tree $\mathcal{T}_\nu$ on their y-segments; at each node μ in $\mathcal{T}_\nu$ we have a sorted list $\mathcal{L}_\mu$ on the z-coordinates of the rectangles in the canonical subset of node μ. To speed up the search in these lists we apply fractional cascading. A query in $\mathcal{T}_\nu$ is performed as follows. We search in $\mathcal{T}_\nu$ with p_y, the y-coordinate of the starting point of the ray. For each node μ on the search path, we search in $O(\log\log n)$ time with p_z in the list $\mathcal{L}_\mu$; this gives us the rectangle in $S(\mu)$ with largest z-coordinate smaller than p_z, which is $\Phi_\rho(S(\mu))$. This way the planar problem can be solved with $O(\log n \log\log n)$ query and update time, using $O(n \log n)$ storage. We obtain the following theorem.

Theorem 6.9 *Ray shooting queries into a fixed direction in a set of axis-parallel polyhedra with n vertices in total can be answered in $O(\log^2 n \log\log n)$ time with a structure that uses $O(n \log^2 n)$ storage. An axis-parallel polyhedron of constant complexity can be inserted into or deleted from the structure in $O(\log^2 n \log\log n)$ amortized time.*

6.2 c-Oriented Polyhedra

In this section we present a number of structures for ray shooting into a fixed direction in a set of c-oriented polyhedra. Without loss of generality, we assume that this fixed direction is the negative z-direction. First, we employ the general strategy of Chapter 4. This leads to a structure with $O(c^2 \log n (\log\log n)^2)$ query time and $O(n \log n)$ storage. The technique of Section 4.2 to reduce the query time can also be used, yielding $O(c^2 \log n)$ query time with $O(n^{1+\epsilon})$ storage. Then we present a structure whose query time is only linear in c, namely $O(c \log^2 n)$, and that uses $O(n \log^2 n)$ storage. This structure only works for non-intersecting polyhedra. Finally, we present a dynamic structure with $O(c^2 \log^2 n \log\log n)$ query time and $O(\log^2 n \log\log n)$ update time, using $O(n \log^2 n)$ storage.

6.2.1 Using the General Strategy

Before we apply the general strategy, we decompose the faces of the polyhedra into quadrilaterals as described in Section 4.3.2. Now consider a set S of n quadrilaterals such that the top edges of the quadrilaterals are parallel to each other, the bottom edges of the quadrilaterals are parallel to each other, and the left and right edges of the quadrilaterals are parallel to each other. Assume without loss of generality that the left and right edges of the quadrilaterals in S are parallel to the y-axis, and that the bottom edges are parallel to the x-axis. We split the quadrilaterals in S into a rectangular part and a triangular part with a segment that is parallel to the x-axis. The set S_R of rectangular parts can be handled in the same way the axis-parallel polyhedra were handled in the Section 6.1.1. It remains to find a structure for ray shooting in the set S_T of triangular parts. This structure is based on the strategy described in Section 4.1.

Recall that to apply Theorem 4.2 we need (i) a $(1/r)$-cutting for the triangles in S_T, and (ii) a structure for line intersection queries. Since the triangles in S_T are parallel to the xy-plane, we can obtain a $(1/r)$-cutting of size $O(r)$ in the same way as in Section 6.1.2: sort the triangles on their z-coordinates and take a plane through every $\lfloor n/r \rfloor$th triangle. See Lemma 6.4. In particular, we can obtain a $(1/2)$-cutting of size 2. (More precisely, the cutting consists of two half-spaces that contain $\lfloor |S_T|/2 \rfloor$ and $\lceil |S_T|/2 \rceil$ triangles, respectively.) It remains to devise a structure for line intersection queries.

Because the direction of the query ray is fixed to be the negative z-direction, we will only perform line intersection queries with lines that are parallel to the z-axis. Notice that a vertical line l intersects a triangle if and only if the intersection point of l with the xy-plane is contained in the projection of the triangle onto the xy-plane. Hence, line intersection queries in a set S' of triangles can be answered by a point location query in the union of the projections of the triangles: line l intersects at least one triangle if and only if the intersection point of l with the xy-plane is contained in the union of the projected triangles. We call this union the *shadow* of S' and denote it by $\mathcal{SH}(S')$. Normally this approach does not lead to efficient solutions, as the shadow of a set of triangles can have quadratic complexity. In our case, however, the triangles are *homothets*, that is, they are identical up to scaling and translation. In this case the complexity of the shadow is only linear [9]. The fact that the triangles are homothets also has another pleasant consequence, namely that we can use the fast point location scheme of [45]. With this scheme, a point location query takes only $O((\log \log n)^2)$ time, provided that the coordinates of the query point are normalized. Fortunately we have to do this normalization only once, at the start of the query algorithm. After that we can answer the line intersection queries in $Q_I(n) = O((\log \log n)^2)$ time.

This concludes the description of the structure for the set S_T of triangular parts. As remarked before, the set S_R of rectangular parts is handled as in Section 6.1.1. This leads to the following theorem.

Theorem 6.10 *Ray shooting queries into a fixed direction in a set of c-oriented polyhedra with n vertices in total can be answered in $O(c^2 \log n (\log \log n)^2)$ time with a structure that uses $O(n \log n)$ storage. The structure can be built in $O(n \log^2 n)$ randomized time.*

Proof: Let us analyze the performance of the structure for S_T. The bounds on the structure for the rectangular pieces follows from Theorem 6.3. We have shown above that intersection queries can be answered in $O((\log \log n)^2)$ time after an initial cost of $O(\log n)$ for normalization. According to Theorem 4.2 the total query time is $O(\log n(\log \log n)^2)$.

The shadow of a set of homothetic triangles has linear complexity [9], so the point location structure that answers line intersection queries uses $M_I(n) = O(n)$ storage and $O(n \log n)$ randomized preprocessing time [45]. Recall that we have a $(1/2)$-cutting of size 2. Hence, the storage requirement $M(n)$ of the total structure satisfies $M(n) = O(n) + 2M(n/2)$, which solves to $O(n \log n)$. To prove the bound on the preprocessing time we have to be a bit careful. Except for the $O(n \log n)$ time that we need to construct the point location structure for the shadow, we also need to construct the shadow itself. Note that the shadow to be stored at a node is the union of the shadows of its two children. The union of two shadows of total size n can be computed in $O(n \log n)$ time by a standard sweep line algorithm, because we know that the resulting shadow has linear complexity. Hence, we can construct the ray shooting structure in a bottom-up fashion in time $T(n)$, where $T(n) = O(n \log n) + 2T(n/2)$. This solves to $T(n) = O(n \log^2 n)$. The bounds now follow from Lemma 4.7. $\square$

6.2.2 Reducing the Query Time

Achieving a query time of $O(c^2 \log n)$ is fairly straightforward, given the results of the previous sections. First, we decompose the faces of the polyhedra into quadrilaterals as described in Section 4.3.2. The resulting set of quadrilaterals is partitioned into $O(c^2)$ subsets according to their type. This time, however, we do not split the quadrilaterals into a rectangular and a triangular part, but we treat them as a whole. We give a brief description of the structure that stores one of the subsets. The structure is based on the general strategy of Section 4.2, and it is very similar to the structure of Section 6.1.2. Because the quadrilaterals are parallel to each other we can construct a $(1/r)$-cutting of size $O(r)$. Furthermore, an intersection sequence query can be answered by a binary search. These two ingredients are exactly the same as in Section 6.1.2. Also the third ingredient, a structure for priority intersection queries, is obtained in a similar manner: we use a structure from Chapter 5. In particular, we apply Theorem 5.7. Note that the partitioning of the set of quadrilaterals into $2c(c-1)$ subsets of different types already has taken place, when we perform a priority intersection query. Hence, priority intersection queries take $O(\log n)$ time, and not $O(c^2 \log n)$ time as Theorem 5.7 suggests. With these observations the next theorem is straightforward.

Theorem 6.11 *For any $\varepsilon > 0$, there exists a structure for ray shooting into a fixed direction in a set of c-oriented polyhedra with n vertices in total that has $O(c^2 \log n)$ query time and uses $O(n^{1+\varepsilon})$ storage and preprocessing time.*

6.2.3 Reducing the Dependency on c

The query time of the data structures for ray shooting in a c-oriented set of polyhedra that we presented in the two previous sections depends quadratically on c. In this section we show that it is possible to reduce the dependency on c to linear. The dependency on n increases slightly, resulting in a query time of $O(c \log^2 n)$. Thus the solution of this section is superior to the solutions of the previous sections when $c = \Omega(\log n)$. Note that in that case the query time is even better than the query time of the structure presented in Section 5.2.1, where we studied the easier problem of ray shooting from a fixed point. However, the new solution only works for a set of non-intersecting polyhedra.

Let us go back to the decomposition scheme described in Section 4.3.2. Recall that this scheme results in a set of quadrilaterals, each of which has two edges that are parallel to the yz-plane—called the right edge and the left edge—, and a top and a bottom edge that each have one of c possible orientations. We then proceeded by partitioning the set of quadrilaterals into $2c(c-1)$ subsets, according to the orientation of their top and bottom edges. Here we use the same decomposition into quadrilaterals, but the partitioning into subsets is different: because we want to have a linear dependency on c, we cannot afford to distinguish according to the orientation of both top and bottom edges. So we put two quadrilaterals in the same subset if and only if their bottom edges are parallel. Observe that the quadrilaterals in one subset are not necessarily parallel to each other. We will store each of the c resulting subsets of quadrilaterals in a separate structure. Next we describe the structure for such a subset S.

The starting point of our study is the general strategy of Section 4.1. Thus we would like to find a cutting for the quadrilaterals in S, and we would like to have an efficient structure for line intersection queries. Due to our partitioning scheme we know that all the bottom edges of the quadrilaterals in S are parallel to each other. Unfortunately, this is not enough to ensure the existence of a cutting of small size. Moreover, the shadow (the union of the projections onto the xy-plane) of a subset of S does not necessarily have small complexity, which would be useful for obtaining an efficient structure for line intersection queries. Both problems disappear if we first store the quadrilaterals in a segment tree; we will show that the desired properties hold for the canonical subsets that are stored at the nodes in the segment tree.

To simplify the notation, let us assume that the common orientation of the bottom edges is parallel to the x–axis. Recall that the decomposition into quadrilaterals was done by cutting the faces with edges that are parallel to the yz-plane. Hence, each quadrilateral has a well defined x-interval. We store the quadrilaterals in a segment tree $\mathcal{T}$ according to these x-intervals.

Consider a node ν in $\mathcal{T}$. Let $S(\nu)$ be its canonical subset. As usual, we restrict our attention to the parts of the quadrilaterals whose x-coordinate is in I_ν, the interval corresponding to ν. It will be useful to partition $S(\nu)$ into two subsets: a subset $S^+(\nu)$ of quadrilaterals whose projected top edges have positive slope, and a subset $S^-(\nu)$ of quadrilaterals whose projected top edges have non-positive slope. In other words, if the length of the right edge of a quadrilateral is greater then the length of its left edge then we put the quadrilateral in $S^+(\nu)$, otherwise we put it in $S^-(\nu)$. Consider the set $S^+(\nu)$; the set $S^-(\nu)$ can be handled analogously. $S^+(\nu)$ has two important properties, which we will prove below. First, the complexity of the shadow of (any subset of) $S^+(\nu)$ is linear, and second, there is no cyclic overlap among the quadrilaterals in $S^+(\nu)$. The first fact implies that an efficient structure for line intersection queries exists, and the second fact enables us to impose an order on the quadrilaterals. Therefore we can apply the general strategy of Section 4.1 to compute $\Phi_\rho(S^+(\nu))$. Next we discuss these two properties in more detail.

A line intersection query with a line l in a subset $S' \subseteq S^+(\nu)$ asks whether at least one of the quadrilaterals in S' is intersected by l. Because we will only query with lines that are parallel to the z-axis, this query can be answered by determining whether the intersection of l with the xy-plane lies in $\mathcal{SH}(S')$, the shadow of S'. Fortunately, this shadow has small complexity.

Lemma 6.12 *The complexity of $\mathcal{SH}(S')$ is $O(|S'|)$.*

Proof: Consider the complement of $\mathcal{SH}(S')$. This complement consists of cells. Since the quadrilaterals have a bottom side that is parallel to the x–axis and a top side whose projection has positive slope, each cell is bounded to the left by the left boundary of the slab and not by edges of quadrilaterals. See Figure 6.2. Hence, the

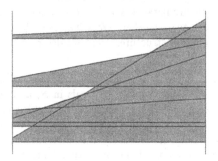

Figure 6.2: The shadow of a set $S' \subset S^+(\nu)$.

total complexity of the boundaries of all cells is no more than the maximal complexity of the zone of a line in an arrangement of lines. It is well known that this complexity is linear. See [50].   □

We can use a standard point location method to determine whether a point is inside $\mathcal{SH}(S')$ or not, but due to the special nature of the shadow this is not necessary. The fact that the bottom edges of the quadrilaterals are parallel to the x-axis and that the top edges have positive slope implies that every line parallel to the x-axis intersects the boundary of the shadow only once. See Figure 6.2. In other words, if we partition $\mathcal{SH}(S')$ into *strips*—by drawing lines parallel to the x-axis through the vertices of the boundary of the shadow—then inside a strip we can decide in constant time whether the query point lies in $\mathcal{SH}(S')$. Hence, we store the vertices of $\mathcal{SH}(S')$ in a list $\mathcal{L}$ that is sorted on y-coordinate. A line intersection query with a vertical line l is performed as follows: First, we find the strip that contains the intersection of l with the xy-plane by searching with the y-coordinate of l in $\mathcal{L}$; then we decide in constant time whether the intersection of l with the xy-plane lies inside $\mathcal{SH}(S')$.

We have seen how to answer line intersection queries. The second ingredient for the general strategy of Section 4.1 is a cutting. Recall that this cutting is used to impose an order on the objects. Here we do not have to use cuttings, since we can obtain the desired order directly. For two quadrilaterals f and f', we define $f \prec f'$ if and only if a ray into the negative z-direction exists that first intersects f and then intersects f'. An order that is consistent with $\prec$ (that is, an order $f_1, f_2, \ldots$ on the quadrilaterals in $S(\nu)$ such that $f_i \prec f_j$ implies $i < j$) is called a *depth order* for $S(\nu)$ (in the negative z-direction). Clearly, such an order is exactly what we need in order to answer a ray shooting query using binary search. Depth orders are studied extensively in Part C of this thesis. To prove that $S^+(\nu)$ admits a depth order, we make the following definitions. Let $I_\nu = [x_\nu : x'_\nu]$ be the x-interval corresponding to ν, and define h'_ν to be the vertical plane $x = x'_\nu$. Finally, let $\tilde{S}^+(\nu) = \{f \cap h'_\nu | f \in S^+(\nu)\}$ be the set of intersections of the quadrilaterals in $S^+(\nu)$ with the plane h'_ν. Note that these segments are non-intersecting, because the quadrilaterals do not intersect. Since any set of non-intersecting segments in the plane admits a depth order [124], the following lemma establishes the existence of a depth order on $S^+(\nu)$.

Lemma 6.13 *A depth order on $\tilde{S}^+(\nu)$ corresponds to a depth order on $S^+(\nu)$.*

Proof: If a ray ρ intersects a quadrilateral f_i then the projection of ρ onto the plane h'_ν will also intersect $f_i \cap h'_\nu$, since the bottom edges of the quadrilaterals in $S^+(\nu)$ are parallel to the x-axis and the projected top edges have positive slope. Now suppose that ρ first intersects a quadrilateral f_i and later a quadrilateral f_j. Then the projection of ρ onto h'_ν will intersect both $f_i \cap h'_\nu$ and $f_j \cap h'_\nu$. Moreover, they are intersected in the same order since f_i and f_j do not intersect. Hence, $f_i \prec f_j$ implies $f_i \cap h'_\nu \prec f_j \cap h'_\nu$, and a depth order on $\tilde{S}^+(\nu)$ corresponds to a depth order on $S^+(\nu)$.   □

This leads to the following data structure for computing $\Phi_\rho(S^+(\nu))$. The quadrilaterals in $S^+(\nu)$ are stored in the leaves of a binary search tree $\mathcal{T}_\nu$ according to the depth order. At each node μ in $\mathcal{T}_\nu$ there is an associated structure for line intersection queries on the set $S^+(\mu)$ of the quadrilaterals that are stored in the subtree rooted at μ. This associated structure is a sorted list $\mathcal{L}_\mu$ on the y-coordinates of the vertices of $\mathcal{SH}(S^+(\mu))$. To facilitate the search in these lists we apply fractional cascading.

How do we search in this tree? When we search with a ray ρ starting above all faces, then things are not too complicated: we just turn left at nodes μ where $l(\rho)$ intersects at least one of the quadrilaterals in $S^+(lchild(\mu))$, and we turn right otherwise. But when ρ does not start above all faces then we cannot replace ρ by $l(\rho)$. To perform the search correctly, we have to know the position of the starting point p of ρ in the depth order on $S^+(\nu)$. The faces that come before p in the ordering are, by definition, never intersected by ρ, and the faces that come after p are intersected by ρ if and only if they are intersected by $l(\rho)$. If we know the position of p in the depth order on $S^+(\nu)$ then it is easy to restrict our attention to the faces which come after p in the order. Recall that the depth order on $S^+(\nu)$ corresponds to a depth order on the set $\tilde{S}^+(\nu)$. In the same way we can argue that the position of p in the depth order on $S^+(\nu)$ corresponds to the position of $\tilde{p}$, the orthogonal projection of p

onto the plane h'_ν, in the depth order on the set $\tilde{S}^+(\nu)$. The latter position is easily determined by a point location query with $\tilde{p}$ in $\mathcal{V}(\tilde{S}^+(\nu))$, the vertical adjacency map of $\tilde{S}^+(\nu)$.

We finished the description of the structure to compute $\Phi_\rho(S^+(\nu))$, so now we can give an overview of the total structure for computing $\Phi_\rho(S)$.

- The main tree is a segment tree $\mathcal{T}$ on the x-intervals of the quadrilaterals.

  - At every node ν in $\mathcal{T}$ there is a balanced binary tree $\mathcal{T}^+_\nu$ for the set $S^+(\nu)$. The order in which the quadrilaterals are stored in these trees is a depth order into negative z-direction.

    * At every node μ in T_ν, the vertices of $\mathcal{SH}(S^+(\mu))$ are stored in a list $\mathcal{L}_\mu$ sorted on y–coordinate. With each vertex we store the boundary segment of $\mathcal{SH}(S^+(\mu))$ that crosses the strip below it.

  - At every node ν in $\mathcal{T}$ there is a similar tree $\mathcal{T}^-_\nu$ for the set $S^-(\nu)$.

  - Fractional cascading is applied to trees $\mathcal{T}^+_\nu$ and $\mathcal{T}^-_\nu$ with their associated lists.

  - At every node ν in $\mathcal{T}$ there are point location structures for $\mathcal{V}(\tilde{S}^+(\nu))$ and $\mathcal{V}(\tilde{S}^-(\nu))$.

We briefly summarize the query algorithm for this structure.

Algorithm 6.14
Input: The query ray ρ.
Output: $\Phi_\rho(S)$.
1. Search with p_x in $\mathcal{T}$.
2. **for** each node ν on the search path
3.   **do** Locate $\tilde{p}$ in $\mathcal{V}(\tilde{S}^+(\nu))$.
4.   Compute $f^+_\nu = \Phi_\rho(S^+(\nu))$ by searching in $\mathcal{T}^+_\nu$.
5.   Locate $\tilde{p}$ in $\mathcal{V}(\tilde{S}^-(\nu))$.
6.   Compute $f^-_\nu = \Phi_\rho(S^-(\nu))$ by searching in $\mathcal{T}^-_\nu$.
7. **return** the nearest of the faces f^+_ν and f^-_ν found in lines 4 and 6.

Finally, let us describe how to build this structure. The first step is to construct the segment tree $\mathcal{T}$ and insert the x-intervals of the quadrilaterals. Now we know for each node ν its canonical subset $S(\nu)$. Next we describe how the associated structure $\mathcal{T}^+_\nu$ that stores $S^+(\nu)$ is constructed; $\mathcal{T}^-_\nu$ is constructed in a similar fashion. First, we compute a depth order on the set $S^+(\nu)$ by computing a depth order on $\tilde{S}^+(\nu)$. Then we apply the following recursive procedure. Split $S^+(\nu)$ into two subsets $L^+(\nu)$ and $R^+(\nu)$ such that $L^+(\nu)$ contains the first half of the quadrilaterals in the depth order and $R^+(\nu)$ contains the second half. Recursively build structures for these two subsets. The structure for $L^+(\nu)$ will be the left subtree of the total structure, and the structure for $R^+(\nu)$ will be the right subtree. It remains to construct the list $\mathcal{L}_{root(\mathcal{T}^+_\nu)}$ of the vertices of $\mathcal{SH}(S^+(\nu))$. This list can be computed from the lists that we already computed for $L^+(\nu)$ and $R^+(\nu)$ by a simple merge. The merging is performed by walking simultaneously along the two lists, adding the vertices that are encountered to $\mathcal{L}_{root(\mathcal{T}_\nu^+)}$ and also adding intersection points between boundary segments stored in these two lists. The segments to be stored with the vertices in $\mathcal{L}_{root(\mathcal{T}_\nu^+)}$ can be computed during the process. When all this has been done, we build the planar point location structures and the fractional cascading structure.

Recall that we actually have c structures like the one we just described, one for each subset S that we created in partitioning scheme described at the beginning of this subsection. A query is performed in each structure, and of the at most c faces that are found the one closest to p is selected. This leads to the following theorem.

Theorem 6.15 *Ray shooting queries into a fixed direction in a set of c-oriented non-intersecting polyhedra with n vertices in total can be answered in $O(c \log^2 n)$ time with a structure that uses $O(n \log^2 n)$ storage. The structure can be built in $O(n \log^2 n)$ time.*

Proof: We show that the structure that stores a subset S of quadrilaterals reports the first face in S that is hit in $O(\log^2 |S|)$ time, and that it uses $O(|S| \log^2 |S|)$ storage and preprocessing time. From this the theorem readily follows, since we have c such subsets whose total size is $O(n)$.

If we can prove that at a node ν in $\mathcal{T}$ we can report the first face that is hit in $O(\log |S(\nu)|)$ time, and that the associated structures use $O(|S(\nu)| \log |S(\nu)|)$ storage and preprocessing time, then the bounds follow by standard arguments using the properties of segment trees. Without loss of generality, we consider the associated structure for the set $S^+(\nu)$ of some node ν in $\mathcal{T}$. The correctness of our approach follows from the correctness of the general strategy of Section 4.1, and Lemma 6.13. To prove the bound on the preprocessing time, we note that the computation of the depth order on $\tilde{S}^+(\nu)$ can be done in $O(|S^+(\nu)| \log |S^+(\nu)|)$ time. See Chapter 10 of Part C. From Lemma 6.12 it follows that the merging step to compute $\mathcal{L}_\mu$ from $\mathcal{L}_{lchild(\mu)}$ and $\mathcal{L}_{rchild(\mu)}$ takes linear time. Hence, the recursive procedure for constructing a tree $\mathcal{T}_\nu^+$ takes $O(|S^+(\nu)| \log |S^+(\nu)|)$ time. The planar point location structure for $\mathcal{V}(\tilde{S}^+(\nu))$ can be constructed in the same amount of time, and the application of fractional cascading does not influence the time bounds. The bound on the amount of storage and the preprocessing time follows. As for the query time, we note that the point location takes $O(\log |S^+(\nu)|)$ time, and the search in a tree $\mathcal{T}_\nu$ takes the same amount of time, because the fractional cascading allows us to search in a shadow in constant time. □

6.2.4 Dynamization

A dynamic structure for ray shooting into a fixed direction in a set of c-oriented polyhedra can be obtained by making a slight modification to the structure of Section 5.2.2, which we developed for dynamic ray shooting from a fixed point. We briefly describe the resulting structure.

As usual, we decompose the faces into quadrilaterals of $O(c^2)$ different types, and we use a separate structure for each type. See Section 4.3.2. Consider a set of quadrilaterals of a fixed type, and assume without loss of generality that the left and right side of the quadrilateral are parallel to the y-axis. As in Section 5.2.2, we build a segment tree $\mathcal{T}$ on the x-segments to reduce the dimension of the problem by one. At each node ν in $\mathcal{T}$, we split the quadrilaterals into a rectangular part and a triangular part. The rectangular pieces can be handled using another segment tree $\mathcal{T}_\nu$; this is the same as in Section 6.1.3. It remains to handle the triangular pieces. In Section 5.2.1 we showed that the triangular pieces can be handled by storing a set $V_T^-(\nu)$ of points in the plane, such that the point with largest second coordinate of all the points whose first coordinate lies in a certain range can be reported. We were looking for the point with largest second coordinate, because we were shooting from a fixed point with z-coordinate greater than the z-coordinates of all the faces. (The faces with greater z-coordinate could be handled separately.) However, when we shoot into a fixed direction instead of from a fixed point, we are searching for the point in $V_T^-(\nu)$ with largest second coordinate that is smaller than p_z. This means that we can no longer use priority search trees. Instead, we use a two-dimensional range tree. A two-dimensional range tree uses $O(n \log n)$ storage, and the query and update times are $O(\log n \log \log n)$ if we apply dynamic fractional cascading, see [86]. We arrive at the following theorem.

Theorem 6.16 *Ray shooting queries into a fixed direction in a set of c-oriented polyhedra with n vertices in total can be answered in $O(c^2 \log^2 n \log \log n)$ time with a structure that uses $O(n \log^2 n)$ storage. A c-oriented polyhedron of constant complexity can be inserted into or deleted from the structure in $O(\log^2 n \log \log n)$ amortized time.*

6.3   Axis-Parallel and c-Oriented Curtains

Recall that a curtain is an unbounded polygon in 3-space with two edges that are parallel to the z-axis and extend to minus infinity. We show that ray shooting into a fixed direction in a set of axis-parallel or c-oriented curtains can be done slightly more efficient than if we are dealing with arbitrary axis-parallel or c-oriented polyhedra. This will be useful in Part D to obtain a more efficient hidden surface removal algorithm.

6.3.1   A Static Structure

Let S be a set of n axis-parallel curtains. Split S into two subsets, a subset S_1 containing the curtains with top edges parallel to the x-axis, and a subset S_2 with top edges parallel to the y-axis. We describe a structure for ray shooting in S_1; the structure for S_2 is similar. To simplify the notation and without loss of generality, we may assume that the fixed direction of the query ray is the positive y-direction. Again we employ the general strategy of Section 4.1. Because the curtains in S_1 are all parallel to the xz-plane, a $(1/2)$-cutting of size 2 is easily found: we just take a plane that is parallel to the xz-plane such that half of the curtains lies on either side. This leads to a binary search tree $\mathcal{T}$ on the y-coordinates of the curtains; at each node

ν in $\mathcal{T}$ there is a structure for line intersection queries in the set $S_1(\nu)$ of curtains whose z-coordinate is stored in the subtree rooted at ν.

How does this associated structure look like? Because the query lines have a fixed direction—they are parallel to the y-axis—the following approach can be used. A query line l intersects at least one curtain if and only if the intersection of l with the xz-plane is contained in the shadow of the curtains onto the xz-plane, that is, the union of the orthogonal projections of the curtains onto the xz-plane. Thus we only have to preprocess this shadow for point location. Observe that the shadow of a set of curtains is exactly the area below the projections of the top edges onto the xz-plane. Hence, the upper envelope (see Section 5.1.1) of these projected top edges is the boundary of the shadow. See Figure 6.3. We store the vertices of the upper

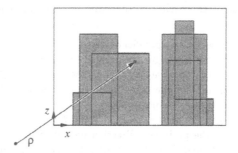

Figure 6.3: Answering line intersection queries by searching in the upper envelope.

envelope of the projected boundary edges of the curtains in $S_1(\nu)$ in a list $\mathcal{L}_\nu$ that is sorted on x-coordinate. A line intersection query can be answered in constant time after we have located the x-coordinate of the query line in this list. Since we have to search many lists, we apply fractional cascading to speed up the search. Thus a line intersection query can be answered in $Q_l(n) = O(1)$ time, after an initial search that takes $O(\log n)$ time.

The structure described above enables us to compute $\Phi_\rho(S_1)$; a similar structure can be used to compute $\Phi_\rho(S_2)$. This leads to the following theorem.

Theorem 6.17 *Ray shooting queries into a fixed direction in a set of n axis-parallel curtains can be answered in $O(\log n)$ time with a structure that uses $O(n \log n)$ storage. The structure can be built in $O(n \log n)$ time.*

Proof: According to Theorem 4.2, the total query time is $O(\log n + Q_l(n)\log n) = O(\log n)$, because the line intersection queries can be answered in constant time after an initial cost of $O(\log n)$. The bound on the amount of storage and the preprocessing time follows, because we can compute the lists $\mathcal{L}_\nu$ in linear time using presorting. See Section 5.1.1. □

Trivially, we can extend this approach to c-oriented curtains by partitioning the set of curtains into c subsets: two curtains are in the same subset if and only if their top edges are parallel. This leads to the following theorem.

Theorem 6.18 *Ray shooting queries into a fixed direction in a c-oriented set of n curtains can be answered in $O(c \log n)$ time with a structure that uses $O(n \log n)$ storage. The structure can be built in $O(n \log n)$ time.*

6.3.2   Dynamization

If we want to dynamize the structure of the previous subsection, we need to maintain the shadow of a set of curtains dynamically. However, it is difficult to maintain upper envelopes efficiently in a dynamic environment. Our dynamic data structure therefore uses a different approach than the static structure presented in the previous subsection.

Let S be a dynamic set of axis-parallel curtains. We describe the structure for the subset S_1 of curtains whose top edges are parallel to the x-axis, the structure for the subset S_2 of curtains with top edges parallel to the y-axis being similar. As in the previous section, we assume for the sake of exposition that the fixed direction of the query rays is the positive y-direction.

A curtain $f \in S_1$ can be written as $[f_x : f'_x] \times f_y \times [-\infty : f_z]$. A query ray ρ starting at point p into the positive y-direction intersects the curtain f if and only if $p_x \in [f_x : f'_x]$, $p_y < f_y$ and $p_z < f_z$. Clearly, $\Phi_\rho(S_1)$ is the curtain with smallest y-coordinate satisfying these conditions. The structure consists of two levels. The first level selects all the curtains satisfying $p_x \in [f_x : f'_x]$; the second level selects the curtain with smallest y-coordinate such that $p_y < f_y$ and $p_z < f_z$. More precisely, we store the curtains in a segment tree $\mathcal{T}$ according to their x-intervals. Let $S_1(\nu)$ be the canonical subset of node ν. At each node ν we store the set $\{(f_y, f_z) : f \in S_1(\nu)\}$ in a priority search tree $\mathcal{T}_\nu$. This a dynamic structure that stores a set of points in the plane and makes it possible to find the leftmost point in a half-infinite vertical slab. A ray shooting query is performed as follows. We search with p_x in $\mathcal{T}$, and for every node ν on the search path we compute $\Phi_\rho(S_1(\nu))$ by querying the priority search tree $\mathcal{T}_\nu$ with the half-infinite vertical slab $[p_y : \infty] \times [p_z : \infty]$. Of the $O(\log n)$ curtains that we find we select the one with smallest y-coordinate. We obtain the following theorem.

Theorem 6.19 *Ray shooting queries into a fixed direction in a set of n axis-parallel curtains can be answered in $O(\log^2 n)$ time with a structure that uses $O(n \log n)$ storage. An axis-parallel curtain can be inserted into or deleted from the structure in $O(\log^2 n)$ time.*

Proof: Priority search trees use a linear amount of storage and have query and update times of $O(\log n)$. This fact, together with Theorem 2.2, proves the theorem. $\square$

The method can be generalized to c-oriented curtains in the same straightforward way as in the static case.

Theorem 6.20 *Ray shooting queries into a fixed direction in a c-oriented set of n curtains can be answered in $O(c \log^2 n)$ time with a structure that uses $O(n \log n)$ storage. A c-oriented curtain can be inserted into or deleted from the structure in $O(\log^2 n)$ time.*

6.4 Arbitrary Polyhedra

Ray shooting into a fixed direction in a set of arbitrary polyhedra can be solved in a manner that is similar to the solution that we presented in Section 5.3.2. We give a brief description of the method.

By Lemma 4.8 we can restrict our attention to shooting in a set S of n triangles. The strategy we adopt is the same as in Section 5.3.2: first, we select all triangles that are intersected by $l(\rho)$, the line through the query ray, and then we shoot in the arrangement of planes through the selected triangles. The selection of the intersected triangles in a number of canonical subsets can be done using a three-level partition tree, as in Section 5.3.2. (However, this time the projection of the triangles onto the xy-plane is a projection into the fixed direction of the query rays.) Let $S(\xi)$ be a canonical subset of a third-level tree, and let $H_{S(\xi)}$ be the set of planes containing the triangles in $S(\xi)$. The way $H_{S(\xi)}$ is processed is different from Section 5.3.2: since we are not shooting from a fixed point we cannot restrict our attention to a single cell, and we have to preprocess the full arrangement $\mathcal{A}(H_{S(\xi)})$ for ray shooting queries.

Lemma 6.21 *There exists a structure for ray shooting with arbitrary rays in an arrangement $\mathcal{A}(H)$ of n planes in $\mathbb{E}^3$ that has $O(\log n)$ query time and uses $O(n^3)$ storage and preprocessing time. It is also possible to obtain, for any $\varepsilon > 0$ and any $n \leqslant m \leqslant n^3$, $O(n^{1+\varepsilon}/m^{1/3})$ query time with a structure that uses $O(m^{1+\varepsilon})$ storage and preprocessing time. The latter structure is dynamic; a plane can be inserted or deleted in $O(m^{1+\varepsilon}/n)$ amortized time.*

Proof: First, we describe a structure with $O(\log n)$ query time that uses $O(n^3)$ storage. We build in $O(n^3)$ time a point location structure for $\mathcal{A}(H)$, using Chazelle's method [18]. Next, we explicitly construct the full arrangement $\mathcal{A}(H)$ in $O(n^3)$ time [50]. Each cell in the arrangement is preprocessed in linear time (linear in the size of the cell) for $O(\log n)$ time ray shooting queries, using the hierarchical representation of Dobkin and Kirkpatrick [48]. A ray shooting query in the set H now proceeds as follows. First we search in $O(\log n)$ time with the starting point of the ray in the point location structure. Then a ray shooting query is performed in the structure that is associated with the leaf in the point location structure, also taking $O(\log n)$ time. This gives us $\Phi_\rho(H)$. Observe that the hierarchical representation of Dobkin and Kirkpatrick [48] allows ray shooting into any direction. Hence, the solution described above for ray shooting in an arrangement works for arbitrary rays.

Using almost the same approach we can obtain a trade-off between the amount of storage and the query time. See also [6]. We describe a structure for rays that are directed downward; upward directed rays can be handled in a similar manner. For downward directed rays, we first select all the planes that lie below the starting point of the ray, using a partition tree. If we only consider this subset of planes, then we can restrict our attention to the cell of the (sub)arrangement that is the intersection of the positive half-spaces bounded by the planes. This convex polytope is preprocessed for ray shooting queries using the dynamic structure described by Agarwal and Matoušek [3]. Their structure uses $O(n^{1+\varepsilon})$ storage, it has $O(\log^3 n)$ query time, and $O(n^\varepsilon)$ amortized update time. (In fact, they also describe a dynamic structure for ray shooting in an arrangement of planes with the same performance as the structure described here.) With a partition tree that uses $O(m^{1+\varepsilon})$ storage and has $O(m^{1+\varepsilon})$ update time, we can select the planes below a query point in $O(n^{1+\varepsilon}/m^{1/3})$ canonical subsets. The bounds follow. (See Theorem 2.15.)   □

Let us briefly summarize the overall approach that we take. First, we select the triangles intersected by $l(\rho)$ in a small number of canonical subsets, using a three-level partition tree. For such a canonical subset, the problem has been reduced to ray shooting in the arrangement of planes through the triangles in the subset. This problem can be solved using Lemma 6.21. Because all triangles in $S(\xi)$ are intersected by $l(\rho)$, the triangle contained in $\Phi_\rho(H_{S(\xi)})$ is $\Phi_\rho(S(\xi))$. We conclude with the following theorem. The proof of this theorem is fairly standard, and we leave it as an (easy) exercise to the reader.

Theorem 6.22 *For any $\varepsilon > 0$, there exist a structure for ray shooting into a fixed direction in a set of arbitrary polyhedra with n vertices in total that has $O(\log n)$ query time and uses $O(n^{3+\varepsilon})$ storage and preprocessing time. It is also possible to obtain, for any $n \leqslant m \leqslant n^3$, $O(n^{1+\varepsilon}/m^{1/3})$ query time with a structure that uses $O(m^{1+\varepsilon})$ storage and preprocessing time. The latter structure is dynamic; a polyhedron of constant complexity can be inserted or deleted in $O(m^{1+\varepsilon}/n)$ amortized time.*

Remark 6.23 If the polyhedra are disjoint then one can obtain a structure that has $O(\log n)$ query time and uses only $O(n^{2+\varepsilon})$ storage. We leave this as an exercise to the reader.

Chapter 7

Ray Shooting with Arbitrary Rays

After having studied instances of the ray shooting problem where either the starting point of the query ray or its direction is fixed, the time has come to tackle the general problem. All the structures that we present in this chapter are based on the general strategy of Chapter 4. The tools that we need to obtain the various ingredients which are needed in the general strategy include cuttings, random sampling and Plücker coordinates.

7.1 Axis-Parallel and c-Oriented Polyhedra

In this section we study the problem of ray shooting with arbitrary rays in a set of axis-parallel polyhedra. Using the general strategy of Section 4.2 we are able to obtain a structure that has $O(\log n)$ query time and uses $O(n^{2+\varepsilon})$ storage and preprocessing time, for any fixed $\varepsilon > 0$. In the next subsection we present the global structure, analyze its performance and state the final result. The global structure uses an associated structure that answers priority intersection queries, which we describe in a separate subsection. But before we do this, we study so-called intersection counting queries: Count the number of faces in a set of axis-parallel polyhedra which are intersected by a query line. The study of these queries, which are interesting in their own right, will give us some ideas on how to answer priority intersection queries. In the last subsection of this section we extend the solution to c-oriented polyhedra. This results in a data structure that uses $O(n^{2+\varepsilon})$ storage and has a query time of $O(c^2 \log n)$.

7.1.1 The Global Structure

Consider a set of axis-parallel polyhedra. As usual, we decompose the faces into axis-parallel rectangles, and we partition the resulting set of rectangles into three subsets according to their orientation. By Lemma 4.6 we can restrict our attention to the set S of axis-parallel rectangles that are parallel to the xy-plane.

The first ingredient we need to apply the general strategy of Section 4.2 is a $(1/r)$-cutting for S, where $r = n^{\varepsilon_1}$. Lemma 6.4 guarantees that we can construct in $O(n \log n)$ time a $(1/r)$-cutting of size $O(r)$: since S consists of rectangles that are parallel to the xy-plane, we can sort the rectangles on their z-coordinate and take a plane through every $\lfloor n/r \rfloor$th rectangle. By Lemma 6.5—which was also valid for arbitrary rays—we can answer intersection sequence queries in this cutting in $O(\log r)$ with a structure that uses $O(r)$ storage and that can be built in $O(r \log r)$ time. To apply Theorem 4.4 we furthermore need a structure for priority intersection queries. We shall see in Section 7.1.3 that these queries can be answered in $O(\log n)$ time after $O(n^{2+\varepsilon_2})$ preprocessing, for any $\varepsilon_2 > 0$. This leads to the following theorem.

Theorem 7.1 *For any $\varepsilon > 0$, there exists a structure for ray shooting in a set of axis-parallel polyhedra with n vertices in total that has $O(\log n)$ query time and uses $O(n^{2+\varepsilon})$ storage and preprocessing time.*

Proof: The query time follows immediately from Theorem 4.4 and Lemma's 6.4, 6.5 and 7.6. Because n_S, the number of possible suffixes, is $O(r)$, the preprocessing time $T(n)$ satisfies

$$T(n) = O(n \log n) + O(r)O(n^{2+\varepsilon_2}) + O(r \log r) + O(r)T(n/r),$$

where we can choose $\varepsilon_2 > 0$ arbitrarily small. Recall that $r = n^{\varepsilon_1}$. It follows from the First Recurrence Lemma that, for any $\varepsilon > 0$, we can choose $\varepsilon_1 > 0$ and $\varepsilon_2 > 0$ such that the recurrence solves to $O(n^{2+\varepsilon})$. Obviously, the storage requirement is bounded by the same quantity.   □

In the introduction to this part it was mentioned that Schmitt et al. [111] have presented a structure with $O(n^{0.695})$ query time that uses $O(n \log^{O(1)} n)$ storage. This structure is based on the conjugation tree of Edelsbrunner and Welzl [55]. If we use Matoušek's partition trees instead, we obtain a structure whose query time is only $O(n^{1/2+\varepsilon})$. Moreover, it is possible to trade storage for query time. These results can also be obtained by adapting the structure that we present in this section. See also [102]. For completeness we state the result as a theorem.

Theorem 7.2 *For any $\varepsilon > 0$ and any m with $n \leqslant m \leqslant n^2$, there exists a structure for ray shooting in a set of axis-parallel polyhedra with n vertices in total that has $O(n^{1+\varepsilon}/\sqrt{m})$ query time and uses $O(m^{1+\varepsilon})$ storage and preprocessing time.*

7.1.2 Intersection Counting Queries

Before we study priority intersection queries, we first describe a structure for *intersection counting queries* in a set S of axis-parallel rectangles that are parallel to the xy-plane. An intersection counting query in S asks for the number $\sigma_l(S)$ of rectangles in S that are intersected by a query line l. The method that we present naturally extends to counting the number of faces in a set of axis-parallel polyhedra that are intersected by a query line, by the usual decomposition scheme; the number of rectangles of the decomposition that are intersected is the same as the number of faces that are intersected.

The planar case

Let us first study the planar version of intersection counting queries. Thus we have a set S of line segments in, say, the xz-plane. We do not require the segments to be axis-parallel, although this will be the case when we extend the solution to three-dimensional space. Given a query line l, which also lies in the xz-plane, we want to compute $\sigma_l(S)$, the number of segments in S that are intersected by l. Recall from Section 2.7 that a segment s is intersected by a line l if and only if l^*, the point that is dual to l, is contained in s^*, the double wedge that is dual to s. Hence, an intersection counting query can be answered by locating l^* in the arrangement $\mathcal{A}(S^*)$ of lines that bound the dual wedges of the segments in S. This arrangement can be constructed incrementally in $O(n^2)$ time [50]. For a cell c of $\mathcal{A}(S^*)$, we define $\sigma_c(S)$ to be the number of segments that are intersected by lines l with $l^* \in c$. Observe that $\sigma_c(S)$ can be computed from $\sigma_{c_1}(S)$ in constant time if c and c_1 share an edge. Therefore we can compute the numbers $\sigma_c(S)$ for all cells c in $\mathcal{A}(S^*)$ in $O(n^2)$ time in total, by traversing the arrangement in a standard manner. Preprocessing the arrangement for $O(\log n)$ time point location takes $O(n^2)$ time, using Kirkpatrick's method [73] for example.

Lemma 7.3 *Intersection counting queries in a set of n segments in the plane can be answered in $O(\log n)$ time with a structure that uses $O(n^2)$ storage and preprocessing time.*

The three-dimensional case

We now turn our attention to the three-dimensional case, where we are given a set S of axis-parallel rectangles that are parallel to the xy-plane. For a rectangle $f \in S$, let $\tilde{f}$ be the orthogonal projection of f onto the xz-plane, and let $\hat{f}$ be the orthogonal projection of f onto the yz-plane. Observe that $\tilde{f}$ and $\hat{f}$ are axis-parallel segments. For a line l, let $\tilde{l}$ and $\hat{l}$ be defined analogously. The following observation, which is easy to prove, is crucial in our solution: a rectangle $f \in S$ is intersected by a line l if and only if $\tilde{f}$ is intersected by $\tilde{l}$ and $\hat{f}$ is intersected by $\hat{l}$. The structure for intersection counting queries will be a two-level structure. The first level selects, in a small number of canonical subsets, all rectangles f such that $\hat{f} \cap \hat{l} \neq \varnothing$. The second level is used to count for each selected canonical subset the number of rectangles f with $\tilde{f} \cap \tilde{l} \neq \varnothing$. This can be done efficiently by Lemma 7.3.

We continue to describe this two-level structure more carefully. Let L be the set of lines in the yz-plane that are the duals of the endpoints of the segments in $\hat{S} = \{\hat{f} : f \in S\}$. In other words, L is the set of lines bounding the double wedges that are dual to the segments in $\hat{S}$. Construct a $(1/r)$-cutting $\Xi(L)$ (see Section 2.5) of size $O(r^2)$ for L, where r is a parameter to be determined later. Let t be a triangle in this cutting and let $\hat{l}$ be a line in the yz-plane such that $\hat{l}^*$, the dual of $\hat{l}$, lies in t. Consider a segment $\hat{f} \in \hat{S}$. Either $t \subset \hat{f}^*$ and we are certain that $\hat{l}$ intersects $\hat{f}$, or $t \cap \hat{f}^* = \varnothing$ and we know that $\hat{l}$ misses $\hat{f}$, or one or both of the lines bounding $\hat{f}^*$ intersects t. In the latter case we cannot tell whether $\hat{l}$ intersects $\hat{f}$, but fortunately this can happen for only few segments $\hat{f}$: since $\Xi(L)$ is a $(1/r)$-cutting there are at most n/r lines in L that intersect t. Let us define $\hat{S}^{\odot}(t)$ to be the set of segments

$\hat{f}$ whose duals fully contain t, and let $\hat{S}^\times(t)$ be the set of at most n/r segments $\hat{f}$ whose duals partially cover t. For a face $\hat{f} \in \hat{S}^\odot(t)$ and a line l in $\mathbb{E}^3$ such that $\hat{l}^* \in t$, we know that l intersects f if and only if $\tilde{l}$ intersects $\tilde{f}$. Hence, we store the set $\tilde{S}^\odot(t) = \{\tilde{f} : \hat{f} \in \hat{S}^\odot(t)\}$ in the planar intersection counting structure of Lemma 7.3. Of the segments in $\hat{S}^\times(t)$ we do not know (yet) whether they are intersected by $\tilde{l}$, so we store them in a recursively defined structure. The data structure for intersection counting queries in S can thus be described as follows.

- If the number of rectangles in S is some small enough constant then the structure consists of one leaf, where we store the set S explicitly.

- Otherwise, the structure is a tree $\mathcal{T}$ of branching degree $O(r)$. There is a one-to-one correspondence between the children of $root(\mathcal{T})$ and the triangles of a $(1/r)$-cutting $\Xi(L)$ for the set L of lines bounding the double wedges that are the duals of the segments in $\hat{S}$.

  - At $root(\mathcal{T})$ the cutting $\Xi(L)$ is stored, preprocessed for point location.
  - At $root(\mathcal{T})$ we also store for each triangle $t \in \Xi(L)$ a planar intersection counting structure $\mathcal{D}_{IC}(t)$ on $\tilde{S}^\odot(t)$, which is the structure of Lemma 7.3.

- The child ν_t of $root(\mathcal{T})$ corresponding to triangle t is the root of a recursively defined structure on the set $S(\nu_t) = \{f : \hat{f} \in \hat{S}^\times(t)\}$.

An intersection counting query with a line l is performed by a call to the following algorithm with $\nu = root(\mathcal{T})$.

Algorithm 7.4
Input: A node ν in $\mathcal{T}$ and the query line l.
Output: $\sigma_l(S(\nu))$.
1. **if** ν is a leaf
2.   **then** Compute $k = \sigma_l(S(\nu))$ by testing all rectangles in S. **return** k.
3.   **else** Locate $\hat{l}^*$ in $\Xi(L)$. Let t be the triangle containing $\hat{l}^*$.
4.   Compute $k_1 = \sigma_{\tilde{l}}(\tilde{S}^\odot(t))$ using $\mathcal{D}_{IC}(t)$.
5.   Compute $k_2 = \sigma_l(S(\nu_t))$ recursively.
6.   **return** $k_1 + k_2$.

We obtain the following result.

Theorem 7.5 *For any $\varepsilon > 0$, there exists a structure for intersection counting queries in a set of axis-parallel faces with n vertices in total that has $O(\log n)$ query time and uses $O(n^{2+\varepsilon})$ storage and preprocessing time.*

Proof: The correctness of the approach follows from the discussion above. It remains to analyze the query time and the preprocessing. Because the point location in Step 3 can be done in $O(\log r)$ time [73], and the query time of the planar intersection counting structure is $O(\log n)$ according to Lemma 7.3, the total query time $Q(n)$ satisfies

$$Q(n) = O(\log r) + O(\log n) + Q(n/r).$$

For each of the $O(r^2)$ triangles of $\Xi(L)$ we have a structure that uses $O(n^2)$ preprocessing time. This dominates the preprocessing time $T(n)$, which therefore satisfies

$$S(n) = O(r^2 n^2) + O(r^2)S(n/r).$$

The bounds on the preprocessing (use the First Recurrence Lemma) and the query time (use the Second Recurrence Lemma) follow if we set $r = n^{\varepsilon'}$ for an $\varepsilon' > 0$ that is sufficiently small with respect to ε. □

7.1.3 Priority Intersection Queries

With the knowledge gained in the previous subsection, devising a structure for priority intersection queries is a fairly straightforward exercise. Let us go back to the planar setting, where we are given sets $S_1, \ldots, S_k$ of line segments in the xz-plane that are parallel to the x-axis. For a query line l, we want to determine the smallest index i such that S_i contains a segment that is intersected by l. Clearly, the arrangement of the lines that are dual to the endpoints of the segments in $S_1 \cup \cdots \cup S_k$ gives us all the necessary information, since a cell in this arrangement uniquely determines which segments are intersected by a line whose dual is in the cell. So we can associate with each cell a unique integer, which is the smallest index i such that there is a segment in S_i that is intersected by the lines whose duals are in this cell. The computation of the indices for the cells can again be done by traversing the arrangement. To this end we maintain a priority queue on the indices i such that there is a segment in S_i that is intersected by the lines whose duals are in the current cell. When we go from one cell to the next, we either insert or delete one index. Hence, the index that is associated with the new cell can be found in $O(\log n)$ time. The construction time for the planar structure thus becomes $O(n^2 \log n)$ in total. Now that we have a solution to the planar version of the problem at hand, we extend it to a solution to the three-dimensional problem in exactly the same way as in the previous subsection. That is, we first select all the rectangles that are intersected in the projection onto the yz-plane in a small number of canonical subsets, and for each canonical subset we solve the planar problem in the projection onto the xz-plane as described above. This leads to the following lemma.

Lemma 7.6 *For any $\varepsilon > 0$, there exists a structure for priority intersection queries in a set of axis-parallel faces with n vertices in total that has $O(\log n)$ query time and uses $O(n^{2+\varepsilon})$ storage and preprocessing time.*

7.1.4 c-Oriented Polyhedra

The ray shooting result for axis-parallel polyhedra can be extended to c-oriented polyhedra without difficulty. By Lemma 4.7 it is sufficient to consider a set S of quadrilaterals whose top edges are parallel to each other, whose bottom edges are parallel to each other and whose left and right edges are parallel to each other.

We know that the quadrilaterals in S are parallel to each other. Hence, the method of Lemma 6.4 gives us a $(1/r)$-cutting of size $O(r)$. Moreover, an intersection sequence query can be answered in $O(\log r)$ time after $O(r \log r)$ preprocessing. It remains to devise a structure for priority intersection queries in S. Using the same ideas as in the axis-parallel case, this is fairly straightforward. Let us briefly sketch the method. To simplify the notation, we assume that the left and right edges are parallel to the y-axis. Let h_t be a plane that is orthogonal to the top edges of the quadrilaterals, and let h_b be a plane orthogonal to the bottom edges. The basic observation is the following. A quadrilateral is the intersection of two half-planes and one strip in $\mathbb{E}^3$; the half-planes are bounded by the lines through the top and the bottom edge (and contain f, of course), and the strip is bounded by the two lines through the left and right edge. Thus a line intersects a quadrilateral if and only if it intersects both half-planes and the strip. To test whether a line l intersects the half-plane bounded by the top edge we use the plane h_t: l intersects the half-plane if and only if the orthogonal projection of l onto h_t intersects the half-line which is the intersection of the half-plane with h_t. In a similar way we can check whether l intersects the half-plane bounded by the line through the bottom edge, and whether l intersects the strip, leading to the following condition. A line l intersects a quadrilateral f if and only if (i) the orthogonal projection of l onto h_t intersects the half-line which is the intersection of h_t and the half-plane bounded by the top edge of f (ii) the orthogonal projection of l onto h_b intersects the half-line which is the intersection of h_b and the half-plane bounded by the bottom edge of f (iii) the orthogonal projection of l onto the xz-plane intersects the segment which is the intersection of the xz-plane and the strip bounded by the lines through the left and right edge of f. We leave the design of a multi-level structure for priority intersection queries that is based on this observation as an exercise. By now the reader should be familiar enough with cuttings and dualization to be able to design a structure with $O(\log n)$ query time that uses $O(n^{2+\varepsilon})$ preprocessing.

We conclude that ray shooting queries in a set S of quadrilaterals of a fixed type can be answered with the same efficiency as ray shooting queries in axis-parallel rectangles. The following theorem now follows from Lemma 4.7.

Theorem 7.7 *For any $\varepsilon > 0$, there exists a structure for ray shooting in a set of c-oriented polyhedra with n vertices in total that has $O(c^2 \log n)$ query time and uses $O(n^{2+\varepsilon})$ storage and preprocessing time.*

7.2   Curtains

In this section we study the ray shooting problem in a set of curtains. Recall that a curtain is an unbounded polygon in 3-space with three edges, two of which are parallel to the z-axis and extend downward to infinity. Curtains turn out to be an extremely useful special class of polyhedra: they play an important role in the depth order and hidden surface removal algorithms to be presented in Parts C and D, and they can also be used to solve ray shooting queries in other classes of objects such as polyhedral terrains and fat horizontal triangles.

Our structure is based on the general strategy of Section 4.2. In the next subsection we give the global structure of the solution and state the final result, which says that ray shooting queries can be answered in $O(\log n)$ time after $O(n^{2+\varepsilon})$ preprocessing. Then we develop the ingredients that are needed in the strategy, namely data structures for intersection sequence queries and for priority intersection queries. We close this section with a subsection on trade-offs and dynamization and a subsection on the applications of curtains to other ray shooting problems.

7.2.1   The Global Structure

Let S be a set of n curtains. The first step in devising a ray shooting structure using Theorem 4.4 is to construct a $(1/r)$-cutting for S, where $r = n^{\varepsilon_1}$. Since curtains are parallel to the z-axis it is possible to obtain a cutting of reasonably small size.

Lemma 7.8 *Let S be a set of n curtains. It is possible to construct a $(1/r)$-cutting of size $O(r^2)$ for S in time $O(nr)$.*

Proof: Let $\overline{S}$ be the set containing the projections of the curtains in S onto the xy-plane. Let $L_{\overline{S}}$ be the set of lines through the segments in $\overline{S}$. By Theorem 2.6 there exists a $(1/r)$-cutting $\Xi(L_{\overline{S}})$ of size $O(r^2)$ for $L_{\overline{S}}$. We extend each triangle $t \in \Xi(L_{\overline{S}})$ into an (unbounded) three-dimensional simplex $s_t = t \times [-\infty : +\infty]$. In other words, s_t is an infinite vertical 'pillar' whose cross section with the xy-plane is t. It is straightforward to check that $\Xi(S) = \{s_t : t \in \Xi(L_{\overline{S}})\}$ is a $(1/r)$-cutting for S of size $O(r^2)$. The construction time for $\Xi(S)$ is dominated by the construction time for $\Xi(L_{\overline{S}})$, which is $O(nr)$. □

The two other ingredients for Theorem 4.4 are discussed in the next two subsections. It is shown that intersection sequence queries and priority intersection queries can be answered in $O(\log n)$ time after $O(n^{2+\varepsilon_2})$ randomized preprocessing, for any $\varepsilon_2 > 0$. Anticipating these results, we state the final result for ray shooting among curtains.

Theorem 7.9 *For any $\varepsilon > 0$, there exists a structure for ray shooting in a set of n curtains that has $O(\log n)$ query time and uses $O(n^{2+\varepsilon})$ storage and randomized preprocessing time.*

Proof: The query time follows immediately from Theorem 4.4 and Lemma's 7.11 and 7.20. In Lemma 7.11 we will show that n_S, the number of possible suffixes, is $O(r^6)$. Hence, the preprocessing time $T(n)$ satisfies

$$T(n) = O(nr) + O(r^6)O(n^{2+\varepsilon_2}) + O(n^{2+\varepsilon_2}) + O(r^2)T(n/r).$$

Recall that we have taken $r = n^{\varepsilon_1}$. By the First Recurrence Lemma, for any $\varepsilon > 0$ we can choose $\varepsilon_1 > 0$ and $\varepsilon_2 > 0$ such that the recurrence solves to $O(n^{2+\varepsilon})$. □

7.2.2   Intersection Sequence Queries

In this subsection we devise a structure for intersection sequence queries in the cutting $\Xi(S)$. Recall that $\Xi(S)$ is the vertical extension of a two-dimensional cutting $\Xi(L_{\overline{S}})$, where $L_{\overline{S}}$ is the set of lines through the projected curtains. This implies that the intersection sequence of ρ in $\Xi(S)$ exactly corresponds to the intersection sequence of $\overline{\rho}$, the projection of ρ onto the xy-plane, in $\Xi(L_{\overline{S}})$. So we describe a structure for the planar problem.

Let $E = E(\Xi(L_{\overline{S}}))$ be the set of edges of $\Xi(L_{\overline{S}})$. Our data structure first determines all the edges in E that are intersected by $l(\overline{p})$, the line through $\overline{p}$. This, of course, also determines the triangles of $\Xi(L_{\overline{S}})$ that are intersected. Then we do a binary search with the starting point of $\overline{p}$ to determine the starting triangle of $\overline{p}$. Given this starting triangle and the triangles that are intersected by $\overline{p}$, there are only two possibilities left for the intersection sequence, depending on the direction of $\overline{p}$. Deciding between these two possibilities can be done in constant time.

There is one subtlety involved in this scheme. In general, the fact that a line l stabs a set of segments in the plane (that is, intersects all segments) does not fix the order in which they are intersected. Indeed, there still can be a linear number of different orders. So how can we do a binary search with the starting point? What saves us is the way in which we determine the edges in E that are intersected by $\overline{p}$. This will give us the extra property that is needed to know the order. We determine the intersected edges in E in the same way as in the previous section, namely by locating the point $l(\overline{p})^*$ in the arrangement $\mathcal{A}(E^*)$, where E^* is the set of lines that bound the dual wedges of the edges in E. Why does this give us the extra information that we need? Recall from Section 2.7.1 that the dual e^* of an edge e is a *double* wedge, and that the way in which a directed line l hits e—from below or from above—determines which of the two wedges of e^* contains l^*. Thus a point location with $l(\overline{p})^*$ in $\mathcal{A}(E^*)$ does not only tells us which edges in E are intersected by $l(\overline{p})$, but it also tells us how these edges are intersected. It is not hard to prove that if a ray hits two edges from a fixed side, then the order in which they are intersected is fixed. Hence, the extra information that we get by locating $l(\overline{p})^*$ in $\mathcal{A}(E^*)$ determines the order in which the edges are intersected. So we can find the starting simplex using binary search. These considerations lead to the following data structure for intersection sequence queries in $\Xi(S)$.

- The arrangement $\mathcal{A}(E^*)$ is stored, preprocessed for point location queries.

  - For each cell c in $\mathcal{A}(E^*)$ we store the edges that are intersected by lines l with $l^* \in c$ in a binary search tree $\mathcal{T}_c$. The order in which the edges are stored in $\mathcal{T}_c$ corresponds to the order in which they are intersected by these lines.

The intersection sequence of a query ray ρ is found as follows.

Algorithm 7.10
Input: The query ray ρ.
Output: $\mathcal{I}_\rho(\Xi(S))$.
1. Locate $l(\overline{p})^*$ in $\mathcal{A}(E^*)$. Let c be the cell containing $l(\overline{p})^*$.
2. Search with the starting point of $\overline{p}$ in $\mathcal{T}_c$ to determine the starting simplex of ρ.
3. Decide which of the two remaining possibilities is the suffix of ρ.

Recall that the output of the algorithm is in fact not the whole intersection sequence $\mathcal{I}_\rho(\Xi(S))$, but pointers to the node corresponding to the starting simplex of ρ and to the data structure for priority intersection queries for the suffix of ρ. Note that when we search in $\mathcal{T}_c$ we have to decide upon the relative position of the starting point of $\overline{p}$ and an edge of E. This can be done by testing whether $\overline{p}$ intersects the edge. We thus obtain the following lemma.

Lemma 7.11 *Intersection sequence queries in the cutting $\Xi(S)$ of Lemma 7.8 can be answered in $O(\log r)$ time with a structure that uses $O(r^6)$ storage and can be built in $O(r^6 \log r)$ time. Furthermore, the number of different suffixes in $\Xi(S)$ is $O(r^6)$.*

Proof: The query time and the correctness are clear, so let us concentrate on the preprocessing. The cutting $\Xi(L_{\overline{s}})$ has size $O(r^2)$. Hence, E^* consists of $O(r^2)$ lines and the arrangement $\mathcal{A}(E^*)$ has size $O(r^4)$. For each cell in this arrangement we store a subset of the $O(r^2)$ edges of E in a binary search tree $\mathcal{T}_c$. Thus the structure uses $O(r^6)$ storage, which is also a bound on the number of different suffixes in $\Xi(S)$. The preprocessing time follows if we notice that the arrangement $\mathcal{A}(E^*)$ and the point location structure for it can be constructed in $O(r^4)$ time [50, 73], and that a tree $\mathcal{T}_c$ can be constructed in $O(r^2 \log r)$ time. $\qquad\qquad\square$

7.2.3 Priority Intersection Queries

The only missing ingredient for our ray shooting shooting structure is a structure for priority intersection queries. At the heart of this structure is a solution to the so-called polytope priority searching problem in five-dimensional space, which we describe first. Having this structure available, we can solve the priority intersection problem by adding two extra levels on top of the data structure.

Polytope priority searching

The polytope priority searching problem is defined as follows. We are given an ordered collection of polytopes in $\mathbb{E}^5$, and we want to report the first polytope that does *not* contain a given query point. More precisely, we are given an ordered collection $H_1, H_2, \ldots, H_m$ of m sets of hyperplanes in $\mathbb{E}^5$, with $\sum_{i=1}^m |H_i| = n$. Each set H_i defines a convex polytope $\mathcal{P}(H_i) = \bigcap\{h^+ : h \in H_i\}$.[1] We want to preprocess $H = H_1 \cup \cdots \cup H_m$ such that we can efficiently find the smallest index i such that $q \notin \mathcal{P}(H_i)$ for a query point q.

---
[1] The only reason that we write h^+ for the half-space containing $\mathcal{P}(H_i)$ is notational convenience; we do not assume that h^+ is actually the half-space above h.

Another way of looking at this problem is the following. Let $Compl(\mathcal{P}(H_i)) = \mathbb{E}^5 - \mathcal{P}(H_i)$ be the complement of $\mathcal{P}(H_i)$. Define $A(H_i) = Compl(\mathcal{P}(H_i)) \cap \mathcal{P}(H_{i-1}) \cap \cdots \cap \mathcal{P}(H_1)$. Thus $A(H_i)$ is exactly the region where the answer to a query is i. We also define $A(H_{m+1}) = \mathcal{P}(H_m) \cap \cdots \cap \mathcal{P}(H_1)$ to be the region that is contained in all polygons. See Figure 7.1 for an illustration of these definitions. A polytope

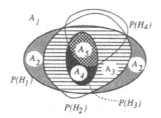

Figure 7.1: The regions $A_i = A(H_i)$ and the subdivision Σ_H they define.

priority query is in fact a point location in the subdivision Σ_H consisting of the regions $A(H_1), \ldots, A(H_{m+1})$ that we just defined. Therefore we need a bound on the complexity of Σ_H, that is, a bound on the total number of features of various dimensions (k-dimensional faces for $1 \leqslant k \leqslant 5$) of Σ_H.

Lemma 7.12 *The complexity of the subdivision Σ_H is $O(n^2 m)$.*

Proof: Note that the number of features of $A(H_i)$ is linear in the number of features of $\mathcal{P}(H_{i-1}) \cap \cdots \cap \mathcal{P}(H_1)$, plus the number of features of $\mathcal{P}(H_i)$, plus the number of features of $\mathcal{P}(H_i) \cap \cdots \cap \mathcal{P}(H_1)$. The complexities of $\mathcal{P}(H_{i-1}) \cap \cdots \cap \mathcal{P}(H_1)$, $\mathcal{P}(H_i)$, and $\mathcal{P}(H_i) \cap \cdots \cap \mathcal{P}(H_1)$ are $O(n^2)$ by the Upper Bound Theorem [50]. Hence, the complexity of $A(H_i)$ is $O(n^2)$, and the bound follows.   □

We now describe a method for point location in Σ_H. The method may not be the most efficient for large values of m, but in our application m is small and we only care about the dependency on n. Our structure is based on random sampling, like Clarkson's point location scheme [34] for arrangements (see Section 2.5). However, due to the fact that our subdivision is not just an arrangement, our structure will be much more involved.

There are two problems in preprocessing Σ_H for point location queries using random sampling. The first problem is that if we take a sample R of the hyperplanes of size r, then the subdivision Σ_R defined by these hyperplanes has size $O(r^2 m)$. If we take r to be a constant, then the $O(m)$ factor dominates the size of the subdivision, resulting in a structure that uses too much storage. This problem is overcome by taking large samples of size $r = n^{1/10}$ so that we can afford the extra $O(m)$ factor in the complexity of Σ_R. (There is nothing magic about the constant $1/10$; it is just a constant that works.) But this imposes a new problem, namely that we can no longer locate the query point in Σ_R in a brute-force way. So we need another structure for locating the query point in Σ_R. Since R is fairly small, we can afford a brutal approach to this problem, which is not possible for Σ_H. A second, and more serious, problem that we encounter is the following. The subdivisions Σ_R that we consider are not full arrangements of hyperplanes. Therefore, random sampling theory does not guarantee anything about the number of hyperplanes that cut a simplex in the triangulated subdivision Σ_R. Moreover, the regions of the subdivision are not convex, so how should we triangulate them? The fact that saves us is that when we consider a region in Σ_R of which we know that it is outside $\mathcal{P}(H_i)$, only the hyperplanes from the sets $H_1, \ldots, H_{i-1}$ are important.

It is time to make these ideas concrete. Let $R \subset H$ be a random sample of size $r = n^{1/10}$, and let $R_i = R \cap H_i$. We extend some definitions that we made for the set H (see Figure 7.1) to the random sample R. Define $\mathcal{P}(R_i) = \bigcap\{h^+ : h \in R_i\}$ and $A(R_i) = Compl(\mathcal{P}(R_i)) \cap \mathcal{P}(R_{i-1}) \cap \cdots \cap \mathcal{P}(R_1)$. If $R_i = \varnothing$ then we define $\mathcal{P}(R_i) = \mathbb{E}^5$; hence, $A(R_i) = \varnothing$ in that case. We also define $A(R_{m+1}) = \mathcal{P}(R_m) \cap \cdots \cap \mathcal{P}(R_1)$. Finally, let Σ_R be the subdivision of $\mathbb{E}^5$ consisting of the regions $A(R_i)$.

Let us first see what we can do about the problem of triangulating Σ_R. As already noted, the regions of Σ_R are not convex and we have no idea how to triangulate them. What we do instead is to construct a set of simplices that may not be a triangulation— in particular, the simplices are not disjoint—but that is good enough for our purposes. We define for each non-empty $A(R_i)$, $1 \leqslant i \leqslant m+1$, a set $Sim(A(R_i))$ of simplices that are disjoint and whose union contains $A(R_i)$. Note that we do not require the union to be equal to $A(R_i)$: the simplices are allowed to extend outside $A(R_i)$. This means that the simplices can intersect simplices from $Sim(A(R_j))$, for some $j \neq i$. The way we define the set $Sim(A(R_i))$ is closely related to the way bottom vertex triangulations are defined (see Section 2.2). Let v be the bottom vertex of $\mathcal{P}(R_i) \cap \cdots \cap \mathcal{P}(R_1)$; if $\mathcal{P}(R_i) \cap \cdots \cap \mathcal{P}(R_1) = \varnothing$ or $i = m+1$, then $A(R_i) = \mathcal{P}(R_{i-1}) \cap \cdots \cap \mathcal{P}(R_1)$ and we choose v to be the bottom vertex of $A(R_i)$. We triangulate the facets of $A(R_i)$ that are not facets of $\mathcal{P}(R_i) \cap \cdots \cap \mathcal{P}(R_1)$, and extend these 4-simplices to 5-simplices, by adding v as a vertex. See Figure 7.2 for an illustration of this definition. Note that

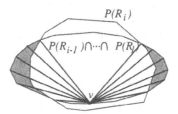

Figure 7.2: The set $Sim(A(R_i))$ of simplices created for the shaded region $A(R_i)$.

$Sim(A(R_i))$ is not well defined for the smallest i such that R_i is non-empty. For this i, we define $Sim(A(R_i))$ to consist of one 'unbounded simplex' $s = \mathbb{E}^5$. A bound on the total number of simplices that we create, which is important for the complexity of our solution, immediately follows from Lemma 7.12.

Lemma 7.13 $\sum_{i=1}^{m+1} |Sim(A(R_i))|$, the total number of simplices, is $O(r^2 m)$.

A property that we need for the correctness of our method is that the union of the simplices constructed for region $A(R_i)$ contains that region.

Lemma 7.14 $A_i \subseteq \bigcup \{s | s \in Sim(A_i)\}$ for every $1 \leqslant i \leqslant m+1$.

Proof: Assume the vertex v that we picked to construct the simplices is the bottom vertex of $\mathcal{P}(R_i) \cap \cdots \cap \mathcal{P}(R_1)$; the case where $\mathcal{P}(R_i) \cap \cdots \cap \mathcal{P}(R_1) = \varnothing$ is proved in a similar way. Let x be an arbitrary point in $A(R_i)$. Shoot a ray from v in the direction of x. After the ray passes through x it will hit a facet f of $A(R_i)$. Since $v \in \mathcal{P}(R_i) \cap \cdots \cap \mathcal{P}(R_1)$, and $A(R_i) \subseteq Compl(\mathcal{P}(R_i))$, this cannot be a facet of $\mathcal{P}(R_i) \cap \cdots \cap \mathcal{P}(R_1)$. Thus x is contained in the 5-simplex that we constructed out of the 4-simplex that we hit in f.   □

The idea of the structure for point location in Σ_R is simple. Suppose that the query point q lies in $A(R_i)$, and that the simplex $s \in Sim(A(R_i))$ contains q. Since the regions $A(R_i)$ are disjoint and the simplices of $Sim(A(R_i))$ are also disjoint, there is only one such simplex; we call it the *interesting simplex* for q. Since $q \in A(R_i)$, we know that $q \notin \mathcal{P}(R_i)$ and, hence, that $q \notin \mathcal{P}(H_i)$. So the only remaining question is whether $q \notin \mathcal{P}(H_j)$ for some $j < i$. This is the case if there is a hyperplane $h \in H_j$ such that $q \in h^-$. Either this hyperplane does not intersect s, in which case all the points in s lie in h^-, or h intersects s. The first case is handled by storing for every simplex $s \in A(R_i)$ the value $i_s = \min(i, \min\{j : s \subset h^- \text{ for some } h \in H_j\})$. The latter case is tested with a recursively defined structure. To keep the size of the structure and the query time bounded, it would be nice if s is intersected by few hyperplanes from $H_1, \ldots, H_{i-1}$. Specifically, we require that our sample R satisfies the following condition:

(C1) Each simplex in $Sim(A(R_i))$, $1 \leqslant i \leqslant m+1$, is intersected by $O(\frac{n}{r} \log r)$ hyperplanes from $H_1 \cup \cdots \cup H_{i-1}$.

We leave the existence of samples with this property for later and continue to describe the data structure. What we have not described so far is how to find the interesting simplex for q. Recall that there can be many other simplices that contain q, but that only one of them is interesting. Let F be the set of $O(r^2 m)$ hyperplanes containing the facets of simplices in $Sim(R) = \bigcup_{i=1}^{m+1} Sim(A(R_i))$. We claim that we can find the interesting simplex for q by locating q in the arrangement $\mathcal{A}(F)$ of hyperplanes.

Lemma 7.15 The cell of $\mathcal{A}(F)$ that contains q uniquely determines the interesting simplex.

Proof: A point $q \in A(R_i)$ is separated from a point $q' \in A(R_j)$ by a hyperplane from R. Since $R \subset F$, this means that the cell of $\mathcal{A}(F)$ containing q determines the region $A(R_i)$ containing q. Similarly, a point $q \in s$ is separated from a point $q' \in s'$, where s and s' are different simplices in $Sim(A(R_i))$, by a plane through a facet of s. This plane is in F and, hence, the cell of $\mathcal{A}(F)$ containing q determines the simplex of $Sim(A(R_i))$ containing q.   □

We are almost ready to give an overview of the data structure. There is one technicality that we must address, however. Recall that we choose $r = n^{1/10}$ so that the $O(m)$ factor does not dominate the $O(r^2 m)$ bound on the number of simplices in $Sim(R)$. But the value of n decreases when we go into recursion, whereas the value of m remains the same. To avoid running into trouble when n gets too small, we stop the recursion when $n = m$—we call a node where this happens a *small node*—and we solve the problem 'brute force' with a point location structure for $\mathcal{A}(H)$ that uses $O(m^{5+\delta})$ preprocessing and has $O(\log n)$ query time. A point location in $\mathcal{A}(H)$ solves the problem, because for each cell c of $\mathcal{A}(H)$, the smallest index i such that $q \notin \mathcal{P}(H_i)$ is the same for every $q \in c$. At this point it is convenient to introduce the notation $\kappa_H(q)$ for the smallest index i such that $q \notin \mathcal{P}(H_i)$, that is, for the answer to a query with point q. We just argued that $\kappa_H(q)$ is the same for every point q in a cell c of $\mathcal{A}(H)$; we denote this index by $\kappa_H(c)$. With this notation, the data structure and query algorithm can be described as follows.

- If the number of hyperplanes in H is less than or equal to m, then we have a small node: we store the full arrangement $\mathcal{A}(H)$, preprocessed for point location queries. For each cell c in $\mathcal{A}(H)$ we store the value $\kappa_H(c)$.

- Otherwise, we have a tree $\mathcal{T}$ of branching degree $O(r^2 m)$. There is a one-to-one correspondence between the children of $root(\mathcal{T})$ and the simplices of $Sim(R)$, where $R \subset H$ is a sample satisfying condition (C1).

  - At $root(\mathcal{T})$ we store a point location structure for the arrangement $\mathcal{A}(F)$, where F is the set of hyperplanes containing the facets of the simplices in $Sim(R)$. For each cell in $\mathcal{A}(F)$ we store the corresponding interesting simplex.
  - For each simplex $s \in Sim(A(R_i))$, $1 \leqslant i \leqslant m + 1$, we store the value $i_s = \min(i, \min\{j : s \subset h^- \text{ for some } h \in H_j\})$.

- The child ν_s of $root(\mathcal{T})$ corresponding to simplex s is the root of a recursively defined structure on the set $H(\nu_s)$ containing the hyperplanes in $H_1 \cup \cdots \cup H_{i_s - 1}$ that intersect s.

A polytope priority query with point q is performed as follows.

Algorithm 7.16
Input: A node ν in $\mathcal{T}$, and the query point q.
Output: $\kappa_{H(\nu)}(q)$.
1. **if** ν is a small node
2.   **then** Compute $i = \kappa_{H(\nu)}(q)$ by locating q in $\mathcal{A}(H(\nu))$. **return** i.
3.   **else** Find the interesting simplex s by locating q in $\mathcal{A}(F)$.
4.   Compute $j = \kappa_{H(\nu_s)}(q)$ recursively.
5.   **return** $\kappa_{H(\nu)}(q) = \min(j, i_s)$.

The efficiency of this method is based upon the existence of samples R with property (C1). Fortunately, there are many such samples.

Lemma 7.17 *Let $R \subset H$ be a random sample of size r. The probability that there is a simplex $s \in Sim(A(R_i))$ that is intersected by more than $O(\frac{n}{r} \log r)$ hyperplanes from $H_1 \cup \cdots \cup H_{i-1}$ is at most $1/2$.*

Proof: This claim follows from general results of Haussler and Welzl [70] on ε-nets, as we show next. (The reader is referred to [70] for an explanation of the terminology used below.) First, we define two range spaces with H (our set of hyperplanes) as the set of elements. The ranges B_1 and B_2 of the two range spaces are defined as follows.

$$B_1 = \{H_1, H_1 \cup H_2, \ldots, H_1 \cup \cdots \cup H_m\}$$
$$B_2 = \{\{h \in H; h \cap s \neq \varnothing\}; s \text{ is a simplex}\}.$$

The range spaces (H, B_1) and (H, B_2) have bounded VC-dimension. Hence, also the range space (H, B_3) has bounded VC-dimension, where $B_3 = \{b_1 \cap b_2; b_1 \in B_1, b_2 \in B_2\}$. Therefore a random sample $R \subset H$ of size r is an ε-net for (H, B_3) with high probability, where $\varepsilon = c \log r / r$ for a suitable constant $c > 0$. Now let R be an ε-net for (H, B_3) and consider some $s \in Sim(A(R_i))$. The set b of hyperplanes of $H_1 \cup \cdots \cup H_{i-1}$ intersecting s is a range in B_3. Since no hyperplane of $R_1 \cup \cdots \cup R_{i-1}$ intersects s, and since R is an ε-net for (H, B_3), the number of hyperplanes in b is at most $\varepsilon n = O(\frac{n}{r} \log r)$. $\square$

This establishes the existence of an efficient structure for polytope priority queries. Next we show how to build it. We take (if the number of hyperplanes is greater than m) a random sample $R \subset H$, and compute the regions $A(R_i)$ and the set $Sim(R)$. We test brute-force if R satisfies condition (C1). If not, we discard R and try again. If the sample is good, we construct the point location structure for $\mathcal{A}(F)$. To compute the interesting simplex that corresponds to a certain cell in $\mathcal{A}(F)$ we proceed as follows. While we build the point location structure for $\mathcal{A}(F)$ according to Theorem 2.7, we can maintain the region of current interest. This way we know for every leaf in the point location structure the interesting region $A(R_i)$. It remains to find the simplex $s \in Sim(A(R_i))$ that contains the cell of $\mathcal{A}(F)$ corresponding to the leaf. This is done by testing every simplex with a point inside the cell. The values i_s and the sets $H(\nu_s)$ can also be computed while constructing the point location structure for $\mathcal{A}(F)$. Finally, we compute the structures for the set $H(\nu_s)$ recursively.

Lemma 7.18 *For any $\varepsilon > 0$, there exists a structure for polytope priority queries in a collection of m polytopes in $\mathbb{E}^5$ defined by n hyperplanes in total that has $O(\log n)$ query time and uses $O(n^{2+\varepsilon} m^{6 + c \log(\frac{\log n}{\log m})})$ storage and randomized preprocessing time, where c is a constant,*

Proof: The correctness of the method follows from the preceding discussion, so let us concentrate on its complexity.

Using condition (C1) and the fact that point location queries in $\mathcal{A}(F)$ take $O(\log n)$ time, we see that the query time $Q(n)$ satisfies

$$Q(n) = O(\log n) + Q(c_1 \frac{n}{r} \log r) \qquad \text{if } n > m$$
$$Q(m) = O(\log n)$$

where c_1 is a constant. Recall that $r = n^{1/10}$. Using the Second Recurrence Lemma, we see that the query time is $O(\log n)$.

The proof that the preprocessing is as claimed is slightly more involved. A sample R is tested for condition (C1) in time $O((r^2m)n)$. Since the probability of success is greater than $1/2$ we expect to find a good sample after a constant number of trials. The point location structure for $\mathcal{A}(F)$ stored at the root of the tree uses $O((r^2m)^{5+\delta})$ storage and preprocessing time, for any fixed $\delta > 0$, and it has a query time of $O(\log n)$. In the same time we can compute the values i_s, and find the hyperplanes on which to recurse for each simplex s. The computation of the interesting simplices for the leaves in the point location structure takes $O((r^2m)^{5+\delta}r^2)$ time in total. Using Lemma 7.13 and making a very conservative estimate, it follows that the storage and randomized preprocessing time $T(n)$ used by our structure satisfies

$$
\begin{aligned}
T(n) &\leqslant c_2(r^2m)^{5+\delta}n + c_2r^2m \times T(c_1\tfrac{n}{r}\log r) \qquad \text{if } n > m \\
T(m) &= O(m^{5+\delta})
\end{aligned}
$$

for some constant c_2. Let $T_d(n)$ be the amount of storage used by a subtree of height d storing n hyperplanes, where the height of a subtree is the largest distance of its root to its small nodes. We claim that $T_d(n) \leqslant c_3m^{5+\delta+d}n^{2+\varepsilon}$ for some constant c_3. For $d = 0$ we have $n = m$, so the claim is true in this case. For $d > 1$ the claim follows by induction, if $\delta < 5\varepsilon$ and r and c_3 are sufficiently large:

$$
\begin{aligned}
T_d(n) &\leqslant c_2(r^2m)^{5+\delta}n + c_2r^2m \times T_{d-1}(c_1\tfrac{n}{r}\log r) \\
&\leqslant c_2(r^2m)^{5+\delta}n + c_2r^2m(c_3m^{5+\delta+d-1}(c_1\tfrac{n}{r}\log r)^{2+\varepsilon}) \\
&= c_3m^{5+\delta+d}n^{2+\varepsilon}\left(\tfrac{c_2r^{10+2\delta}}{c_3n^{1+\varepsilon}m^d} + \tfrac{c_2r^2(c_1\log r)^{2+\varepsilon}}{r^{2+\varepsilon}}\right) \\
&= c_3m^{5+\delta+d}n^{2+\varepsilon}\left(\tfrac{c_2n^{1+\delta/5}}{c_3n^{1+\varepsilon}m^d} + \tfrac{c_2(c_1\log r)^{2+\varepsilon}}{r^{\varepsilon}}\right) \\
&\leqslant c_3m^{5+\delta+d}n^{2+\varepsilon}
\end{aligned}
$$

It is not hard to verify that the height of the complete tree is $O(\log(\tfrac{\log n}{\log m}))$. Hence, $T(n) = T_{c\log(\frac{\log n}{\log m})}(n)$, for some constant c, and the bound follows. $\qquad\square$

Priority intersection queries

The structure for polytope priority queries that we developed above is the heart of our structure for priority intersection queries. Adding two more layers to it in a way to be described next will give us the structure that we seek.

Let $S_1, \ldots, S_k$ be an ordered collection of sets of curtains. We want to find the first subset S_i containing a curtain that intersects a query line. Observe that a line l intersects a curtain if and only if l intersects the curtain in the projection onto the xy-plane and l passes below the line through the top edge of the curtain. This observation leads to the following structure. The first level of the structure is used to select, in a small number of canonical subsets, all curtains that are intersected in the projection. In each selected canonical subset we can regard the top edges of the curtains as full lines. The second level imposes the so-called consistent orientation constraint (see Section 2.7.2) on these lines. Now we can use Plücker coordinates to transform the problem to a polytope priority searching problem in Plücker 5-space, which can be solved using the structure described above.

Let us describe the structure is more detail. The curtains of $S = S_1 \cup \cdots \cup S_k$ that are intersected in the projection are selected in the same way as in Section 7.1.2. So we project the curtains in S onto the xy-plane, resulting in a set $\overline{S}$ of segments, and dualize this set to obtain a set $\overline{S}^*$ of double wedges. Next we construct a $(1/r)$-cutting $\Xi(\overline{S}^*)$ for the lines bounding the double wedges in $\overline{S}^*$. For a triangle t in $\Xi(\overline{S}^*)$, we define $S^\odot(t)$ to be the set of curtains f such that $\overline{f}^*$ contains t, and we define $S^\times(t)$ to be the set of curtains such that $\overline{f}^*$ partially covers t. Hence, for lines l with $\overline{l}^* \in t$, we know that $\overline{l} \cap \overline{f} \neq \varnothing$ for each $f \in S^\odot(t)$; the curtains in $S^\times(t)$ may or may not be intersected by l and need further (recursive) treatment, and the remaining curtains are not intersected by l.

Define $S_i^\odot(t) = S^\odot(t) \cap S_i$, and let $L_i(t)$ be the set of lines through the top edges of the curtains in $S_i^\odot(t)$. We know that a line l with $\overline{l}^* \in t$ misses all curtains in $S_i^\odot(t)$ if and only if l passes above all the lines in $L_i(t)$. Hence, we are interested in the first $L_i(t)$ such that l does *not* pass above all the lines in $L_i(t)$. Recall from Section 2.7.2 that a consistently oriented line l passes above a set $L_i(t)$ of lines if and only if $\pi(l)$, the Plücker point of l, is contained in the Plücker polytope $\mathcal{P}_i(t) = \bigcap\{\varpi(l')^+ : l' \in L_i(t)\}$ corresponding to $L_i(t)$. Thus, if the query line is consistently oriented with respect to all lines in $L(t) = L_1(t) \cup \cdots \cup L_k(t)$, then we can find the first set $L_i(t)$ containing a line passing above l by performing a polytope priority query with $\pi(l)$ in the collection of polytopes $\mathcal{P}_1(t), \ldots, \mathcal{P}_k(t)$. The consistent orientation constraint can be enforced in exactly the same way as in [22]. See Section 2.7.2. This leads to the following data structure for answering priority intersection queries in an ordered collection $S = S_1 \cup \cdots \cup S_m$ of curtains.

- If the number of curtains in S is some small enough constant then the structure consists of a single leaf, where we store the set S explicitly.

- Otherwise, we have a tree $\mathcal{T}$ of branching degree $O(r^2)$. There is a one-to-one correspondence between the children of $root(\mathcal{T})$ and the simplices of a $(1/r)$-cutting $\Xi(\overline{S}^*)$ for the set $\overline{S}^*$ of lines bounding the double wedges that are the duals of the projected curtains.

  - At $root(\mathcal{T})$ the cutting $\Xi(\overline{S}^*)$ is stored, preprocessed for point location.

  - At $root(\mathcal{T})$ we also store for each set $S^\odot(t)$, $t \in \Xi(\overline{S}^*)$, an associated structure $\mathcal{T}_t$. This structure is a two-level structure, whose first level imposes the consistent orientation constraint on the lines, and whose second level is a structure for polytope priority searching.

- The child ν_t of $root(\mathcal{T})$ corresponding to triangle t is the root of a recursively defined structure on the set $S(\nu_t) = S^\times(t)$ of curtains whose dual intersects t.

To be able to give a succinct description of the query algorithm, we introduce the notation $\chi_l(S)$ for the smallest index i such that S_i contains a curtain that intersects a query line l. The following algorithm computes this value. It is first called with $\nu = root(\mathcal{T})$.

Algorithm 7.19
Input: A node ν in $\mathcal{T}$, and the query line l.
Output: $\chi_l(S(\nu))$.
1. **if** ν is a leaf
2.   **then** Compute $i = \chi_l(S(\nu))$ brute-force. **return** i.
3.   **else** Locate $\bar{l}^*$ in $\Xi(\overline{S}^*)$. Let t be the triangle containing $\bar{l}^*$.
4.   Compute $i = \chi_l(S^{\odot}(t))$ by querying $\mathcal{T}_t$.
5.   Compute $j = \chi_l(S(\nu_t))$ recursively.
6.   **return** $\chi_l(S_\nu) = \min(i, j)$.

It follows from Section 7.2.1 that the number m of different subsets in a collection is $O(n^{2\varepsilon_1})$, where we can choose $\varepsilon_1 > 0$ arbitrarily small. This fact simplifies the bound in the following lemma.

Lemma 7.20 *For any $\varepsilon > 0$, there exists a structure for priority intersection queries in a collection of $m = n^{2\varepsilon_1}$ sets of curtains with total size n, where we can choose $\varepsilon_1 > 0$ arbitrarily small, that has $O(\log n)$ query time and uses $O(n^{2+\varepsilon})$ storage and randomized preprocessing time.*

Proof: Recall that the second level—which imposes the consistent orientation constraint—is exactly the same as in [22]. Moreover, it can be shown in a fairly standard manner that this level does not influence the asymptotic bounds of our structure. Hence, in the rest of this proof we simply forget about the second level. Using Lemma 7.18 we now see that query time $Q(n)$ satisfies

$$Q(n) = O(\log n) + Q(n/r),$$

and that the amount of storage and the randomized preprocessing time $T(n)$ satisfies

$$T(n) = O(r^2)O(n^{2+\varepsilon}m^{6+c\log(\frac{\log n}{\log m})}) + O(r^2)T(n/r).$$

The recurrence for the query time solves to $O(\log n)$ by the Second Recurrence Lemma, if we take $r = n^{\varepsilon_2}$ for some constant $\varepsilon_2 > 0$. To prove the bound on $T(n)$, note that $\log(\frac{\log n}{\log m}) = O(\log(1/\varepsilon_1))$ for $m = O(n^{2\varepsilon_1})$. Furthermore $\varepsilon_1 \log(1/\varepsilon_1) \to 0$ when $\varepsilon_1 \to 0$. Recall that we can choose $\varepsilon_1 > 0$ and $\varepsilon_2 > 0$ as small as we like. The bound now follows from the First Recurrence Lemma.   □

7.2.4   Trade-Offs and Dynamization

A trade-off between the amount of storage and the query time for curtains will be needed in Part D to obtain an efficient hidden surface removal algorithm. The structure for ray shooting in a set of curtains that we presented does not lend itself very well for obtaining efficient trade-offs. Fortunately a good trade-off is possible if a different approach is taken, as has been shown by Agarwal and Matoušek [3]. Their structure also allows for updates.

Theorem 7.21 (Agarwal and Matoušek [3]) *For any fixed $\varepsilon > 0$ and any m with $n \leqslant m \leqslant n^2$, there exists a structure for ray shooting in a set of n curtains that has $O(n^{1+\varepsilon}/\sqrt{m})$ query time and uses $O(m^{1+\varepsilon})$ storage. A curtain can be inserted into or deleted from the structure in $O(m^{1+\varepsilon}/n)$ amortized time.*

7.2.5   Applications

In this subsection we give two examples of how to use ray shooting among curtains to solve other ray shooting problems. In the first example—ray shooting in terrains— the reduction to curtains is immediate. In the second example—ray shooting in fat horizontal triangles—we need an extra layer on top of the data structure. We only state the result that is obtained with a logarithmic query time. However, the trade-off and dynamization of the previous subsection can also be achieved.

Terrains

A *polyhedral terrain* is formally defined as the graph of a piecewise linear continuous bivariate real function. Hang a curtain from each edge of the terrain. Chazelle et al. [22] observed that the face of a terrain that is hit by a query ray (starting above the terrain) is uniquely determined by the curtain that is hit and from what side it is hit. Hence, ray shooting queries in a terrain can be answered by shooting in a set of (non-intersecting) curtains. Chazelle et al. showed how to preprocess a set of non-intersecting curtains for $O(\log^2 n)$ time ray shooting queries in $O(n^{2+\varepsilon})$ time. Using Theorem 7.9, we immediately obtain an improved solution for ray shooting in a terrain.

Theorem 7.22 *For any $\varepsilon > 0$, there exists a structure for ray shooting in a polyhedral terrain with n edges in total that has $O(\log n)$ query time and uses $O(n^{2+\varepsilon})$ storage and randomized preprocessing time.*

Fat horizontal triangles

As a second application of curtains we study so-called fat horizontal triangles. We call a triangle *fat* if all its internal angles are greater than some fixed constant θ. Fat *horizontal* triangles, that is, fat triangles that are parallel to the xy-plane, have the following important property.

Observation 7.23 *There exists a set of orientations D of constant size such that the following holds. Let v be any vertex of a fat horizontal triangle t. Then it is possible to split t into two triangles with a segment incident to v whose orientation is in D.*

The size of D is inversely proportional to the minimum angle θ of the triangles. For example, we can take the set $D = \{i\theta/2 : 0 \leqslant i < 4\pi/\theta\}$. Let S be a set of n fat horizontal triangles. The property stated above enables us to decompose each triangle $t \in S$ into at most four triangles t_1, t_2, t_3 and t_4, such that each t_i has two edges whose orientations are in D: Pick any vertex of t and split t according to Observation 7.23 using some segment s. Split the two resulting triangles from the vertices opposite s.

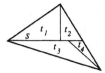

Figure 7.3: Splitting a fat triangle using segments with orientation in D.

See Figure 7.3. This clearly results in four triangles that each have two edges with orientation in D. We call these edges the *fixed edges* of the triangles. Next we partition the set of $4n$ triangles thus obtained into $|D|^2$ subsets $S_1, \ldots, S_{|D|^2}$: two triangles are in the same subset if and only if the two fixed edges of one triangle are parallel to the two fixed edges of the other triangle. For each S_i a separate structure is built. With these structures we can find $\Phi_\rho(S_1), \ldots, \Phi_\rho(S_{|D|^2})$, from which $\Phi_\rho(S)$ is easily computed in $O(|D|^2)$ time.

Consider one subset S_i. Assume without loss of generality that each triangle $t \in S_i$ has one edge that is parallel to the x-axis, and one edge that is parallel to the y-axis. We partition S_i into two subsets: a subset S_i^+ of triangles that lie above (that is, in the positive y-direction of) the edge that is parallel to the x-axis, and a subset S_i^- of triangles that lie below this edge. Consider the set S_i^+; S_i^- is treated in the same way. For a triangle $t \in S_i^+$, we call the edge that is parallel to the x-axis its *bottom edge*, the edge that is parallel to the y-axis its *vertical edge*, and its third edge, which does not have a fixed orientation, its *top edge*. The idea of the structure is as follows. First we select all triangles t such that $l(\rho)$ passes in the y-direction above the line containing the bottom edge of t. Once we know that $l(\rho)$ passes above the bottom edge of these triangles, we can as well extend them to $y = -\infty$. In other words, we can regard each triangle t as a curtain hanging from its top edge into the negative y-direction (which is the direction of its vertical edge). Thus, if we can find all triangles whose bottom edges pass below a query line efficiently, we can use the structure of Theorem 7.9.

How do we find these triangles quickly, and, equally important, in a small number of groups? Here we use the fact that all bottom edges are parallel to the x-axis. So the idea that was used in Section 7.1.4 applies: we project the set $E = E(S_i^+)$ of bottom edges of the triangles in S_i onto the yz-plane, giving a set $\tilde{E}$ of points. A line l passes above the line containing a bottom edge $e \in E$ if and only if $\tilde{e} \in \tilde{l}^-$, where $\tilde{e}$ and $\tilde{l}$ are the projections of e and l onto the yz-plane, and $\tilde{l}^-$ denotes the half-plane below $\tilde{l}$ (that is, in the negative y-direction of $\tilde{l}$). To find all points $\tilde{e} \in \tilde{l}^-$ for a query line l, we dualize the set of points $\tilde{E}$ and construct a $(1/r)$-cutting $\Xi(\tilde{E}^*)$ for the resulting set $\tilde{E}^*$ of lines. The subdivision $\Xi(\tilde{E}^*)$, preprocessed for point location, is stored at the root of our main structure, which is a tree of branching degree $O(r^2)$. For each triangle $t \in \Xi(\tilde{E}^*)$ we have an associated structure on the set of triangles that correspond to the lines below t, and we recursively store the at most n/r lines that intersect t. The associated structure is a ray shooting structure on the set of curtains hanging from the top edges of the triangles into negative y-direction. Choosing $r = n^{\varepsilon'}$, and using the Second Recurrence Lemma, we see that the total query time remains the same as for curtains, namely $O(\log n)$. Using the fact that the ray shooting structure for curtains uses $O(n^{2+\varepsilon})$ preprocessing, the preprocessing can be done in time $O(n^{2+\varepsilon})$ for any (slightly larger) $\varepsilon > 0$, by the First Recurrence Lemma.

Theorem 7.24 *For any $\varepsilon > 0$, there exists a structure for ray shooting queries in a set of n fat horizontal triangles that has $O(\log n)$ query time and uses $O(n^{2+\varepsilon})$ storage and randomized preprocessing time.*

Remark 7.25 Observe that, using the same techniques, it is possible to obtain an alternative solution for the ray shooting problem in a set of axis-parallel polyhedra. This is true because each face of an axis-parallel polyhedron can be split into rectangles whose edges have a fixed orientation. These rectangles can be treated in the same way as the triangles that have two fixed edges: first select the ones whose bottom edge passes below the query line, and then treat the rectangles as curtains hanging from the top edge into the direction of the—in this case two—vertical edges.

7.3 Arbitrary Polyhedra

Finally, we have come to the ray shooting problem in its most general setting: shooting with an arbitrary ray in an arbitrary set of polyhedra. But there is no reason for panic: this section is relatively simple compared to the section on curtains. The structure that we present is based on Theorem 4.4. A slight adaptation of that scheme is necessary, however, to obtain our result, which is a structure with $O(\log n)$ query time, that uses $O(n^{4+\varepsilon})$ preprocessing.

7.3.1 The Global Structure

Recall from Lemma 4.8 that we can restrict our attention to ray shooting in a set S of n triangles in $\mathbb{E}^3$. We want to apply Theorem 4.4, so we need (i) a $(1/r)$-cutting for S, where $r = n^{\varepsilon_1}$, (ii) to be able to answer intersection sequence queries for this cutting, and (iii) to be able to answer priority intersection queries. A structure for priority intersection queries is given in a separate subsection, so let us concentrate on the first two ingredients.

To obtain a cutting we project the triangles in S onto the xy-plane, and we construct a $(1/r)$-cutting $\Xi(E_{\overline{S}})$ for the set $E_{\overline{S}}$ of edges of the projected triangles. Next we erect a vertical pillar from each two-dimensional simplex of the cutting, resulting in a three-dimensional cutting $\Xi(S)$. This is exactly the same as what we did in Lemma 7.8; thus we already know how to answer intersection sequence queries in this cutting (see Lemma 7.11). Let s be a simplex in $\Xi(S)$. We denote the set of triangles that have an edge that intersects s by $S^{\times}(s)$. As usual, for a triangle t in $S^{\times}(s)$ we only consider $t \cap s$, the part inside s. Note that the set $S^{\times}(s)$ not only contains triangles: since we restrict our attention to the part inside s it contains convex k-gons with $3 \leqslant k \leqslant 6$. But this is not a problem. Although it is in fact unnecessary, we can always decompose the parts into triangles if we want to. Observe that $|S^{\times}(s)| \leqslant n/r$.

However, $\Xi(S)$ is not a $(1/r)$-cutting for S: each three-dimensional simplex in $\Xi(S)$ is intersected by at most n/r edges of triangles, but it can be intersected by many triangles whose projection fully contains the corresponding two-dimensional simplex. Let us denote this set of triangles whose projection fully contains the projection of s by $S^\odot(s)$. Next we show that $S^\odot(s)$ can be handled directly, without going into recursion.

To this end we note that, inside s, the triangles in $S^\odot(s)$ can be treated as full planes. Thus we preprocess the arrangement $\mathcal{A}(H_{S^\odot(s)})$ for fast ray shooting, where $H_{S^\odot(s)}$ is the set of planes containing the triangles in $S^\odot(s)$. We have seen in Lemma 6.21 how to do this: the solution that we gave has $O(\log n)$ query time and uses $O(n^3)$ preprocessing. To compute $\Phi_\rho(S^\odot(s))$, we query this structure with ρ (if necessary, we clip ρ so that it starts in s) to find the first plane that is hit. Then we test if this first intersection lies in s. If so, the ray will also hit the triangle that corresponds to this plane; otherwise none of the triangles in $S^\odot(s)$ is hit.

Summarizing, we obtain the following structure.

- If the number of triangles in S is some small enough constant then the structure consists of a single leaf, where all triangles in S are stored.

- Otherwise, the structure is a tree $\mathcal{T}$ of branching degree $O(r^2)$. There is a one-to-one correspondence between the children of $root(\mathcal{T})$ and the simplices of a cutting $\Xi(S)$, where each simplex is intersected by at most n/r triangle edges.

  - At $root(\mathcal{T})$ we have an associated structure $\mathcal{D}_{IS}$ that can answer intersection sequence queries.
  - At $root(\mathcal{T})$ we also have, for every possible suffix $\mathcal{S} = s_0, \ldots, s_t$, an associated structure $\mathcal{D}_{PI}(\mathcal{S})$ that can answer priority intersection queries on the ordered collection of sets $S^\times(s_0) \cup S^\odot(s_0), \ldots, S^\times(s_t) \cup S^\odot(s_t)$.
  - For every simplex $s \in \Xi(S)$, we have a structure for ray shooting in the arrangement $\mathcal{A}(H_{S^\odot(s)})$.

- The child ν_s of $root(\mathcal{T})$ that corresponds to simplex s is the root of a recursively defined structure on the set $S^\times(s)$.

The query algorithm is very similar to the algorithm of Section 4.2, so we only sketch it. First, we determine the intersection sequence of ρ by querying in $\mathcal{D}_{IS}$. Then we recursively search in the starting simplex s_0. This gives us $t^\times = \Phi_\rho(S^\times(s_0))$. To compute $t^\odot = \Phi_\rho(S^\odot(s_0))$, we query in the arrangement $\mathcal{A}(H_{S^\odot(s_0)})$. (This step was not necessary in Section 4.2.) If not both $t^\times$ and $t^\odot$ are NIL then we compute the closest of $t^\times$ and $t^\odot$ and we are done. Otherwise we perform a priority intersection query to determine in which simplex s_i we have to search for the answer. The recursive search in the corresponding subtree gives us $t^\times = \Phi_\rho(S^\times(s_i))$. Finally, we query in the arrangement $\mathcal{A}(H_{S^\odot(s_i)})$ to compute $t^\odot = \Phi_\rho(S^\odot(s_i))$, and we compute $\Phi_\rho(S(s_i)) = \Phi_\rho(\{t^\times, t^\odot\})$. We obtain the following theorem.

Theorem 7.26 *For any $\varepsilon > 0$, there exists a structure for ray shooting queries in a set of polyhedra with n vertices in total that has $O(\log n)$ query time and uses $O(n^{4+\varepsilon})$ storage and randomized preprocessing time.*

Proof: Intersection sequence queries can be performed in $O(\log n)$ time after $O(n^{2+\varepsilon_2})$ preprocessing by Lemma 7.11, and priority intersection queries take $O(\log n)$ time after $O(n^{4+\varepsilon_2})$ preprocessing, as will be shown in the next subsection. The time needed for ray shooting in arrangements is $O(\log n)$—the same as the time for intersection sequence queries and priority intersection queries. Hence, this does not increase the asymptotic query time, which is $O(\log n)$ by Theorem 4.4. Because n_S, the number of possible suffixes, is $O(r^6)$, the preprocessing time $T(n)$ satisfies

$$T(n) = O(nr) + O(r^6)O(n^{4+\varepsilon_2}) + O(r^2)T(n/r),$$

where $r = n^{\varepsilon_1}$, and we can choose $\varepsilon_1, \varepsilon_2 > 0$ arbitrarily small. By the First Recurrence Lemma, the recurrence solves to $O(n^{2+\varepsilon})$ if we make $\varepsilon_1 > 0$ and $\varepsilon_2 > 0$ small enough with respect to ε. Obviously, the storage requirement is bounded by the same quantity. $\square$

It seems difficult to obtain a trade-off between storage and query time with the above structure. But Agarwal and Matoušek [4] have described a different structure which allows for such trade-offs. For completeness we state their result.

Theorem 7.27 (Agarwal and Matoušek [4]) *For any fixed $\varepsilon > 0$ and any m with $n \leqslant m \leqslant n^4$, there exists a structure for ray shooting in a set of n triangles that has $O(n^{1+\varepsilon}/m^{1/4})$ query time and uses $O(m^{1+\varepsilon})$ storage.*

7.3.2 Priority Intersection Queries

Let S be a collection $S_1, \ldots, S_k$ of sets of triangles in $\mathbb{E}^3$. Given a query line l, we want to compute the smallest index i such that S_i contains a triangle that is intersected by l. Let $H_S = \{\varpi(l') : l' \text{ contains an edge of a triangle in } S\}$ be the set of Plücker planes corresponding to the lines through the edges of the triangles. Recall from Section 2.7.2 that the arrangement $\mathcal{A}(H_S)$ contains all the necessary information. Specifically, a point location query with $\pi(l)$ in $\mathcal{A}(H_S)$ tells us exactly which triangles of S are intersected by l. Using this fact, all triangles in S that are intersected by l can be selected in $O(\log n)$ canonical subsets with a structure that uses $O(n^{4+\varepsilon})$ storage. See Theorem 2.17. Obviously, the answer to the query is uniquely determined if we know which triangles are intersected by l. Hence, we can answer a priority intersection query in $O(\log n)$ time if we precompute for every canonical subset the smallest index i such that S_i contains a triangle in that canonical subset.

Lemma 7.28 *For any $\varepsilon > 0$, there exists a structure for priority intersection queries in a set of n possibly intersecting triangles in $\mathbb{E}^3$ that has $O(\log n)$ query time and uses $O(n^{4+\varepsilon})$ storage and randomized preprocessing time.*

Chapter 8

Conclusions

In this part we have developed efficient data structures for several instances of the ray shooting problem for sets of polyhedra in three-dimensional space. The instances varied in the restrictions put on the query rays and on the polyhedra. In particular, we studied ray emanating from a fixed point, rays with a fixed direction, and arbitrary rays, and we considered axis-parallel, c-oriented and arbitrary polyhedra, and curtains. Below we give an overview of our most important results, and we mention some open problems. Readers who want to know all the results on a specific setting of the problem are referred to the corresponding section of this part.

To shed some more light on the complexity of the ray shooting problem, let us compare the amount of storage that we needed to obtain $O(\log n)$ query time (or $O(c^2 \log n)$ in the case of c-oriented polyhedra) in the various settings. This is summarized in Table 8.1. In this table, n is the total number of vertices of all polyhedra and $\varepsilon > 0$ is a constant that can be made arbitrarily small. Although we studied arbitrary curtains only for arbitrary rays, we have included the result in the table, because curtains are important in several applications.

	from a fixed point	into a fixed direction	arbitrary
axis-parallel polyhedra	$n \log n$	$n^{1+\varepsilon}$	$n^{2+\varepsilon}$
c-oriented polyhedra	$n \log n$	$n^{1+\varepsilon}$	$n^{2+\varepsilon}$
arbitrary curtains	—	—	$n^{2+\varepsilon}$
arbitrary polyhedra	$n^2 \alpha(n)$	$n^{3+\varepsilon}$	$n^{4+\varepsilon}$

Table 8.1: The amount of storage needed to obtain a ray shooting structure with $O(\log n)$ query time (or $O(c^2 \log n)$ query time for c-oriented polyhedra).

The question is, of course, whether these bounds can be improved. Let us discuss the optimality of the structures for arbitrary rays, since this is the most interesting case. Here it is worth noting that the best known bound on the amount of storage needed for logarithmic-time ray shooting queries in a set of axis-parallel line segments in the plane is $O(n^2)$.[1] Although no lower bounds are known it is expected that this bound is optimal up to logarithmic factors. This belief is based on the fact that the number of combinatorially different 'query' lines in a set of n segments in the plane is $\Theta(n^2)$. Here two lines are in the same equivalence class if and only if they intersect exactly the same segments. Our structure for the three-dimensional problems of shooting in axis-parallel polyhedra and in curtains have almost the same efficiency as the planar structure and, hence, we believe that they are close to optimal. The complexity of the general ray shooting problem (arbitrary polyhedra and arbitrary rays) seems to be determined by the number of combinatorially different 'query' lines in a set of n given lines in space, where two lines are in the same equivalence class if and only if one can be moved to the other without crossing any of the given lines. This number is $\Theta(n^4)$ [84], which leads us to suspect that our data structure for the general ray shooting problem is close to optimal.

The above discussion gives rise to two open problems: First, there is the problem of proving non-trivial lower bounds for the ray shooting problem. Second, it would be interesting to remove the ε's from the upper bound. In some cases this is indeed possible, if we use some special properties of the cuttings as constructed by Chazelle [18]. See also Matoušek [81], who shows how to get rid of the ε for certain range searching problems. However, this improvement in the amount of storage seems to entail an increase of the query time by some polylogarithmic factor.

We close this chapter with a discussion of trade-offs between the amount of storage and the query time of the ray shooting structures. As before, we restrict our attention to the case of arbitrary query rays. For axis-parallel polyhedra, we have seen in Section 7.1 that one can achieve $O(n^{1+\varepsilon}/\sqrt{m})$ query time using $O(m^{1+\varepsilon})$ storage. Recall from Section 2.6 that there is a lower bound for trade-offs for range searching in $\mathbb{E}^d$: the query time of a structure that uses $O(m)$ storage is necessarily $\Omega(n/(m^{1/d}\log n))$. The discussion above for the logarithmic query time ray shooting structures hints at the fact that ray shooting in axis-parallel polyhedra in 3-dimensional space actually follows the same pattern as range searching for the case $d = 2$. If this is true, then our trade-off is close to optimal. Recall that also for curtains there is a structure that uses $O(m^{1+\varepsilon})$ storage and answers ray shooting queries in $O(n^{1+\varepsilon}/\sqrt{m})$ time. Again, this fits in nicely with the range searching lower bound scheme for $d = 2$. For arbitrary polyhedra, Agarwal and Matoušek [4] have shown how to obtain $O(n/m^{1/4})$ query time using $O(m^{1+\varepsilon})$ storage. This gives an extra indication that the general ray shooting problem follows the range searching lower bound scheme for $d = 4$.

---

[1]Using recent results [18, 79] on cuttings we can improve this slightly to $O(n^2/\log n)$.

Part C

Computing Depth Orders

Chapter 9

Introduction

As was explained in Chapter 1, a possible way of performing hidden surface removal is the painter's algorithm. The most difficult task of this algorithm lies in computing a depth order on the objects in the scene. A depth order on a set S of objects with respect to a point p is defined as follows. Consider two objects $o, o' \in S$. We say that o *obscures* o' with respect to p if and only if there is a ray starting at p that intersects o in a point q and o' in a point q' such that q is closer to p than q'. We write $o \prec_p o'$. A depth order on S with respect to p is an order $o_1, o_2, \ldots$ on S such that $o_i \prec_p o_j$ implies $i < j$. Note that $\prec_p$ is not a transitive relation, and that a pair of objects is not necessarily comparable under $\prec_p$. Also note that there can be a cycle in the relation $\prec_p$—we say that there is cyclic overlap—in which case a depth order does not exist. By fixing the direction $\vec{d}$ of the rays, instead of their starting point, we can define $o \prec_{\vec{d}} o'$—object o obscures object o' in direction $\vec{d}$—and we can define a depth order on S in direction $\vec{d}$. In three-dimensional space one usually considers depth orders in the negative z-direction. If this is the case, then we use the terms 'obscuring' and 'depth order' without the additive 'in the negative z-direction', and we omit the subscript $\vec{d}$ from $\prec$. Thus o obscures o' is equivalent to o is above o'. Notice that a depth order does not exist if two objects intersect. Hence, we only consider sets of non-intersecting objects in this part.

The problem of computing depth orders is a special case of the so-called *separability problem*: Given a set of objects in some space, separate them by a sequence of motions. During the motions, the objects should not collide with each other. (A collision between two objects occurs when their interiors have a non-empty intersection.) Toussaint [118] gives an extensive survey of such problems. When we restrict the type of motion to translations into a fixed direction $\vec{d}$, then the set can be separated if and only if a depth order in direction $\vec{d}$ exists. Thus computing depth orders is not only important for hidden surface removal, but also for motion planning problems where a robot has to (dis)assemble composite objects. In this context, depth orders are called *translation orders*.

Most research on depth orders has concentrated on the two-dimensional version of the problem. The study of two-dimensional depth orders originated in 1980 when Guibas and Yao [65] (see also [124]) studied depth orders for sets of convex polygons in the plane. They showed that depth orders exist for any set of convex polygons and any direction, and gave an optimal $O(n + m \log m)$ algorithm for computing an order, which was later simplified by Ottmann and Widmayer [94]. Here m is the number of polygons and n is the total number of vertices of all polygons. Since then, their work has been extended in several ways. The first extension concerns arbitrary— that is, not necessarily convex—polygons. Observe that there can be a cycle in the relation $(S, \prec)$ if the polygons are not convex, in which case a depth order does not exist. The algorithm should detect such a cycle if there is one, and output a depth order otherwise. Nurmi [89, 91] was the first to give an algorithm for this problem, achieving a time bound of $O(n \log n)$. Recently, Nussbaum and Sack [92] presented an optimal $O(n + m \log m)$ algorithm. Sack and Toussaint [109] showed how to compute in $O(n \log n)$ time all cycle-free directions—that is, all directions for which a depth order exists—for *two* arbitrary polygons, which was improved to $O(n)$ by Toussaint [119]. (In [109, 119] the cycle-free directions are called the *directions of separability*.) Finally, Dehne and Sack [46] studied many of these problems when preprocessing is allowed: after $O(m^2(C_S(n/m) + \log m))$ time and using $O(m^2)$ storage, they are able to answer all kinds of questions on depth orders. Here each polygon is assumed to have n/m vertices and $C_S(x)$ is the time needed to determine all directions of separability of two polygons with x vertices each. $C_S(x)$ varies between $O(\log x)$ for convex polygons, and $O(x)$ for arbitrary polygons. Although their method is efficient when the number of polygons is small, it becomes very costly when there are many polygons. When all polygons have constant size, for example, their preprocessing takes $O(n^2 \log n)$ time and $O(n^2)$ storage and computing an order for a query direction still takes $O(n^2)$ time.

In Chapter 10 it is shown that any set of polygons can be preprocessed in time $O(n \log n)$ into a data structure of size $O(n)$, such that it is possible to determine, for any query direction d, in time $O(\log n)$ whether there exists a depth order. Moreover, such an order can be computed, if it exists, in $O(m)$ time using a structure of size $O(m)$. This improves the results of Dehne and Sack [46] considerably. We also show that all cycle-free directions can be computed in $O(n \log n)$ time. Finally, we extend the method to polyhedral terrains, which are three-dimensional scenes where the projections of the faces onto the xy-plane do not overlap. These results are taken from [38, 40].

In three-dimensional space, the computation of depth orders turns out to be much more difficult than in the plane. Note that—unlike in the planar case—it is possible that there is cyclic overlap among a set of convex objects. See Figure 9.1. In 1984 Nurmi [89] (see also [91]) gave an algorithm for computing three-dimensional depth orders. For a set of n line segments in $\mathbb{E}^3$, his algorithm runs in time $O(n \log n + i)$, where i is the number of intersections of the segments when they are projected onto the xy-plane. In other words, i is the number of pairs of segments that is comparable under $\prec$. Note that i can be $\Theta(n^2)$. Nurmi also extended his algorithm to polyhedra, achieving a bound of $O((n + i) \log n)$, where n is the total number of vertices of the polyhedra and i is the number of intersections between the projected edges. After that not much progress on the three-dimensional depth order problem was made for several years. In 1990 Chazelle et al. [21] considered the case where the objects are lines in $\mathbb{E}^3$. They noted that a depth order can be obtained by a standard sorting

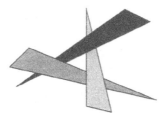

Figure 9.1: Cyclic overlap among triangles.

algorithm, because any two lines can be compared (assuming no two have parallel projections). If there is cyclic overlap, however, then the outcome of the sorting algorithm is not a valid depth order. What makes the problem interesting is that it is not easy to verify in subquadratic time whether a given order on a set of lines is a valid depth order. Chazelle et al. presented an $O(n^{4/3+\varepsilon})$ time algorithm for this problem. However, the problem of computing a depth order for a set of polygons, or even line segments, in space in subquadratic worst-case time remained wide open. Even for the case of axis-parallel segments, it was an open problem to find a depth order in $o(n^2)$ time [106].

In Chapter 11 we settle this open problem by showing that a set of n segments in $\mathbb{E}^3$ can be sorted in subquadratic time. More precisely, we present an algorithm that computes a depth order in $O(n^{4/3+\varepsilon})$ time if it exists, and detects a cycle otherwise. We also show that it is possible to verify a given order in $O(n^{4/3+\varepsilon})$ time. When the segments are c-oriented, that is, they have only c different orientations, then the sorting algorithm runs in $O(cn \log^3 n)$ time and verification takes $O(cn \log^2 n)$ time. Hence, for axis-parallel segments the algorithms run in $O(n \log^3 n)$ and $O(n \log^2 n)$ time, respectively. The algorithms can be extended to handle polygons instead of segments. These results are taken from [44].

Remark 9.1 The painter's algorithm can also be implemented using *binary space partition trees*, called *BSP trees* for short. This method was proposed by Fuchs et al. [59]. The best bound for the size of a BSP tree storing n segments in the plane is $O(n \log n)$, see [100]. Hence, our structure for depth order queries is more efficient. Note that, if the polygons do not allow a depth order, we can always decompose them into a linear number of smaller pieces such that an order exists. In three-dimensional space, binary space partition schemes exist [100] that result in trees of size $O(n^2)$. If there is no cyclic overlap in the query direction then our algorithm for computing depth order is faster. Otherwise, our algorithm only gives a negative result and the BSP tree should be used.

Chapter 10

Depth Orders in the Plane

In this chapter we study depth orders in the plane. More precisely, we study so-called *depth order queries* for a set S of m non-overlapping polygons with n vertices in total. Two polygons are *non-overlapping* if and only if their interiors are disjoint. Thus the boundaries—which we do not consider part of the polygons—are allowed to touch. Depth order queries ask for a depth order on S in a query direction $\vec{d}$. We show that S can be preprocessed into a data structure of size $O(n)$ such that one can decide in $O(\log n)$ time whether a depth order in a query direction exists for S. Moreover, a depth order—if one exists—can be computed in $O(m)$ time with a structure that uses $O(m)$ storage. We also show that all cycle-free directions of S can be computed in $O(n \log n)$ time. These results can be found in Sections 10.3 and 10.4.

It might seem that depth orders in the plane are not very useful in computer graphics; after all, the scenes that one wants to display are usually three-dimensional. However, for an important class of three-dimensional scenes—the so-called polyhedral terrains—solutions to the two-dimensional depth problem can be used. This is shown in Section 10.5. In that section we also adapt the solution to computing depth orders with respect to a query point, instead of in a query direction.

We achieve our results using new results on *relative convex hulls* and *embeddings*, which are of independent interest. The convex hull of a polygon $\mathcal{P}$ relative to a set S of polygons is the polygon that contains $\mathcal{P}$ and excludes S whose boundary has minimal length. This means that the polygon is made 'as convex as possible'; since convex polygons are easier to order than non-convex polygons, this will help us in computing depth orders. We show how to compute in total time $O(n \log n)$ for each polygon in a set S its convex hull relative to the rest of the polygons. This result is discussed in Section 10.1.

An embedding of a set of non-overlapping polygons is a set of non-overlapping polygons such that each polygon in the original set is contained in a unique polygon of the embedding. For convex polygons it is known that there always exist embeddings with a total number of vertices that is linear in the number of polygons [121]. For non-convex polygons this is not always true. However, we show that there exists an embedding with 'few' vertices, which allows us to reduce the amount of storage of the data structure for depth order queries to linear in the number of polygons. In Section 10.2 we show that this embedding can be computed in $O(n \log n)$ time.

10.1   Relative Convex Hulls

In this section we present our results on *relative convex hulls*, a generalization of convex hulls. Relative convex hulls were introduced by Toussaint [119]. They are defined as follows. Define a polygonal circuit to be a closed polygonal path without (proper) self-crossings. Let $\mathcal{P}$ be a polygon and S_1 a set of polygons. The *convex hull of $\mathcal{P}$ relative to S_1*, denoted $\mathcal{CH}(\mathcal{P}|S_1)$, is the polygon whose boundary is the shortest polygonal circuit that includes $\mathcal{P}$ but excludes S_1, that is, with $int(\mathcal{P}) \subseteq int(\mathcal{CH}(\mathcal{P}|S_1))$ and $int(\mathcal{P}') \subseteq ext(\mathcal{CH}(\mathcal{P}|S_1))$ for each $\mathcal{P}' \in S_1$. (Thus these polygons are a slight generalization of simple polygons, where we allow some edges and vertices to be used more than once.) Intuitively, if we release an elastic band that is wrapped around $\mathcal{P}$ then it tries to take the shape of the convex hull of $\mathcal{P}$, but the band can be stopped by the other polygons and it takes the shape of the relative convex hull of $\mathcal{P}$. An

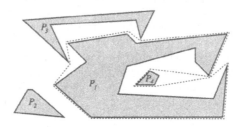

Figure 10.1: A relative convex hull.

example is given in Figure 10.1, where the dashed line is the boundary of the convex hull of $\mathcal{P}_1$ relative to $\{\mathcal{P}_2, \mathcal{P}_3, \mathcal{P}_4\}$. Note that the relative convex hull in Figure 10.1 is not a simple polygon, since there is a vertex that is used twice. Relative convex hulls exhibit the following useful properties:

Lemma 10.1 *Let $\mathcal{P}$ be a polygon and S_1 and S_2 be sets of polygons, such that $\mathcal{P}$ does not intersect any polygon in $S1 \cup S_2$. Then:*
  (i)   *If v is a convex vertex of $\mathcal{CH}(\mathcal{P}|S_1)$, then v is a convex vertex of $\mathcal{P}$.*
  (ii) *If v is a reflex vertex of $\mathcal{CH}(\mathcal{P}|S_1)$, then v is a convex vertex of $\mathcal{P}$ or a convex vertex of some polygon $\mathcal{P}' \in S_1$.*
  (iii) *If $S_1 \subseteq S_2$, then $\mathcal{CH}(\mathcal{P}|S_1) \supseteq \mathcal{CH}(\mathcal{P}|S_2)$.*

Proof: (i): Let v be a convex vertex of $\mathcal{CH}(\mathcal{P}|S_1)$ and let e, e' be the two edges of $\mathcal{CH}(\mathcal{P}|S_1)$ incident on v. If v is a point of $\mathcal{P}$, then v must be a convex vertex of $\mathcal{P}$. If v is not a point of $\mathcal{P}$, then there is a small area around v that does not contain any point of $\mathcal{P}$. More specifically, there are points $q \in e$, $q' \in e'$ ($q, q' \neq v$) such that the triangle determined by q, q' and v does not contain any point of $\mathcal{P}$. But this contradicts $v \in \mathcal{CH}(\mathcal{P}|S_1)$, since replacing $\overline{qv} \cup \overline{vq'}$ by $\overline{qq'}$ yields a polygonal circuit still enclosing $\mathcal{P}$ and excluding S_1 that is shorter.

   (ii): Follows in the same way as (i). Note that if v is a convex vertex of $\mathcal{P}$, then this vertex is used twice by $\mathcal{CH}(\mathcal{P}|S_1)$.

(iii): Let $C_1 = \mathcal{CH}(\mathcal{P}|S_1)$ and $C_2 = \mathcal{CH}(\mathcal{P}|S_2)$. Observe that $\mathcal{P} \subset C_1 \cap C_2$.

First, we show that $C_1 \cap C_2$ is connected. Suppose for a contradiction that $C_1 \cap C_2$ consists of more than one region. Since $\mathcal{P}$ is connected there must be a non-empty connected region $A \subset C_1 \cap C_2$ that is disjoint from $\mathcal{P}$. Since $A \subset C_2$, we also know that A is disjoint from S_2. Region A is bounded by portions $\beta_1, \ldots, \beta_k$ of ∂C_1 and portions $\gamma_1, \ldots, \gamma_k$ of ∂C_2. By (i), these portions cannot have a vertex that is convex with respect to A. Hence, $k \geqslant 2$. But then there must be a segment connecting, say, β_i to β_j (or γ_i to γ_j). This segment lies completely inside C_1 (C_2) and does not intersect $\mathcal{P}$, which contradicts the definition of C_1 (C_2).

Now suppose $S_1 \subseteq S_2$, but $C_1 \not\supseteq C_2$. One easily verifies that it is not possible that $C_1 \subset C_2$. Hence, the boundaries of C_1 and C_2 must intersect. Let β be a maximal portion of ∂C_1 that lies inside C_2. Denote the endpoints of β by p and q, and assume that $int(C_1)$ lies to the left of β when going from p to q. Let $\gamma \subset \partial C_2$ connect p and q such that $int(C_2)$ lies to the right when going from q to p. Because $C_1 \cap C_2$ is connected we can conclude that γ is not intersected by ∂C_1. Denote the area enclosed by β and γ by B. Now β cannot have a vertex that is convex with respect to B. Such a vertex would be reflex with respect to C_1 and thus, by (ii), be a vertex of $\mathcal{P}$ or of a polygon $\mathcal{P}' \in S_1$. The first case cannot occur since it contradicts the fact that C_1 contains $\mathcal{P}$. The second case is impossible since $S_1 \subseteq S_2$, $\beta \subset \partial C_2$ and C_2 excludes S_2. Similarly, γ cannot have a vertex that is convex with respect to B. Such a vertex would be convex with respect to C_2 and therefore be a convex vertex of $\mathcal{P}$. This contradicts the fact that C_1 contains $\mathcal{P}$. Of course, it is impossible that neither γ nor γ' contains a convex vertex and, hence, area B cannot exist.   □

Let S be a set of non-overlapping polygons. For a polygon $\mathcal{P} \in S$, we define $\mathcal{P}^\circ(S) = \mathcal{CH}(\mathcal{P}|S - \{\mathcal{P}\})$ to be the convex hull of $\mathcal{P}$ relative to the rest of S. When the set S is clear from the context, we write $\mathcal{P}^\circ$. Furthermore, we define $S^\circ = \{\mathcal{P}^\circ | \mathcal{P} \in S\}$ to be the set of these relative convex hulls. In the remainder of this section we show that S° can be computed efficiently. Toussaint [119] has shown how to do this for a set of two polygons. We extend his method to larger sets of polygons.

The set S° is computed by Algorithm 10.2 given below. The algorithm roughly works as follows. In lines 1–10, we compute an area around each polygon $\mathcal{P}$, called the *sleeve* of $\mathcal{P}$, that contains the boundary of its relative convex hull. To this end we triangulate the area in between the polygons, and we concatenate the triangles of the triangulation T that share a vertex with $\mathcal{P}$ to each other to form $sleeve(\mathcal{P})$. See Figure 10.2 Then we compute a shortest circuit that starts at a point which we know is on the boundary of the relative convex hull, goes 'around' $\mathcal{P}$, and returns to this point. To compute this circuit we can use an algorithm by Chazelle [15] or by Lee and Preparata [75]. They have shown that if the dual graph of the triangulation of a simple polygon is a chain, then the shortest path between two points in such a polygon can be computed in linear time in the number of vertices of the polygon. (The dual graph is the graph where the nodes are triangles in T and there is an arc between two nodes if the corresponding triangles share an edge.) Thus we must make sure that the dual of $sleeve(\mathcal{P})$ is a chain. Therefore we add in lines 4–5 the triangles of T that are dead-ends to $\mathcal{P}$, so that they will not be added to $sleeve(\mathcal{P})$. In Figure 10.2, for

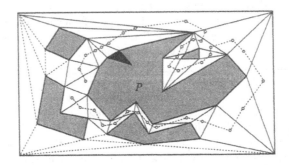

Figure 10.2: The sleeve of $\mathcal{P}$ and its dual chain.

example, the darkly shaded triangle is added to $\mathcal{P}$.

For a polygon $\mathcal{P}$, let $v_0(\mathcal{P})$ be its leftmost vertex. More precisely, $v_0(\mathcal{P})$ is the vertex that is smallest in the lexicographical ordering. Furthermore, let $v_1(\mathcal{P})$ be the next vertex in clockwise order, and let $t_0(\mathcal{P})$ be the triangle in T that shares edge $\overline{v_0(\mathcal{P})v_1(\mathcal{P})}$ with $\mathcal{P}$. A walk along the boundary of $\mathcal{P}$ in clockwise order corresponds to an order on the triangles that share a vertex with $\mathcal{P}$. For a triangle t that shares a vertex with $\mathcal{P}$, the function $next_\mathcal{P}(t)$ returns the next triangle in this order.

Algorithm 10.2
Input: A set S of non-overlapping polygons.
Output: The set S° of relative convex hulls.
1.   Compute a large rectangle r such that $S \subset int(r)$.
2.   Compute a triangulation T of $r - S$, the area inside r in between the polygons.
3.   **for** every $\mathcal{P} \in S$
4.   **do while** there is a triangle $t \in T$ sharing two edges with $\mathcal{P}$
5.   **do** $\mathcal{P} \leftarrow \mathcal{P} \cup t$
6. $t \leftarrow t_0(\mathcal{P})$; $sleeve(\mathcal{P}) \leftarrow \varnothing$;
7.   **repeat**
8.   Concatenate t to $sleeve(\mathcal{P})$
9. $t \leftarrow next_\mathcal{P}(t)$
10.   **until** $t = t_0(\mathcal{P})$
11.   Compute a shortest path from $v_0(\mathcal{P})$ in $t_0(\mathcal{P})$ to the copy of
               v_0 in $t_l(\mathcal{P})$, the last triangle that has been added to $sleeve(\mathcal{P})$.

Theorem 10.3 *Let S be a set of m non-overlapping polygons with n vertices in total. Suppose we can triangulate a rectangular region whose holes are the polygons in S in $T(n, m)$ time. Then the set $S^\circ = \{\mathcal{CH}(\mathcal{P}|S - \{\mathcal{P}\}) : \mathcal{P} \in S\}$ of relative convex hulls can be computed in $O(T(n, m))$ time.*

Proof: First we prove the correctness of the algorithm and then we show that it works within the stated bounds.

It is evident that the circuit that is computed in line 11 of the algorithm contains $\mathcal{P}$ and excludes $S - \{\mathcal{P}\}$. We argue that the boundary of $\mathcal{P}^\circ$ is confined to *sleeve*$(\mathcal{P})$. Suppose that it intersects some triangle $t \in T$ that does not share a vertex with $\mathcal{P}$. Then t has a vertex inside $\mathcal{P}^\circ$ that is not a vertex of $\mathcal{P}$. But then it is a vertex of some other polygon $\mathcal{P}'$ and $\mathcal{P}^\circ$ does not exclude $\mathcal{P}'$. Therefore the boundary of $\mathcal{P}^\circ$ must lie in the union of all triangles that share at least one vertex with $\mathcal{P}$. Furthermore, the triangles that are added to $\mathcal{P}$ in lines 4–5 have only one neighboring triangle. Hence, a circuit around $\mathcal{P}$ that enters t must leave it through the same edge, which results in a useless detour that makes the length of the circuit longer. Hence, these triangles lie completely inside $\mathcal{P}^\circ$, and adding them to $\mathcal{P}$ will not change $\mathcal{P}^\circ$. Finally, since v_0 lies on $\mathcal{CH}(\mathcal{P})$ it will certainly be a vertex of $\mathcal{P}^\circ$. It follows that $\mathcal{P}^\circ$ is indeed the shortest path inside *sleeve*$(\mathcal{P})$ from $v_0(\mathcal{P})$ in $t_0(\mathcal{P})$ to $v_0(\mathcal{P})$ in $t_l(\mathcal{P})$. Because we add in lines 4–5 all the triangles that share two edges with $\mathcal{P}$ to $\mathcal{P}^\circ$, the dual of the triangulation of *sleeve*$(\mathcal{P})$ is a chain. Thus we can use the algorithms of [15] or [75]. This is true even though *sleeve*$(\mathcal{P})$ is not necessarily a simple polygon: some triangle could occur more than once in the sleeve, as happens twice in Figure 10.2. However, Toussaint [119] observed that this is no real problem, because the algorithms in [15, 75] only use 'local' information. Intuitively, we can view *sleeve*$(\mathcal{P})$ as being embedded onto a surface of several levels, so that if a triangle occurs for the second (or third) time it lies 'above' its previous occurrence. Most algorithms that work for simple polygons, in particular the ones that we need, also work in this case.

To prove the time bound, we note that lines 1–2 can be implemented to run in $O(n + T(n,m))) = O(T(n,m))$ time. We may assume that we get the triangulation in the form of its doubly-connected-edge-list (DCEL). With this representation, and knowing that the shortest path algorithms of [15, 75] take linear time, it is relatively straightforward to implement lines 4–11 such that they take $O(|sleeve(\mathcal{P})|)$ time. It follows that the total time taken to execute lines 4–11 for all polygons $\mathcal{P}$ is $\sum_{\mathcal{P} \in S} O(|sleeve(\mathcal{P})|)$. A triangle is added to *sleeve*$(\mathcal{P})$ only if it shares a vertex with $\mathcal{P}$. Hence, any triangle can occur at most three times in a sleeve (that is, once in three sleeves, three times in one sleeve, et cetera) and the total complexity of all sleeves is $O(n)$. The time bound follows, as well as the bound on the amount of storage. $\square$

Any order on S naturally corresponds to a unique order on S°, and vice versa. This correspondence is also preserved when restricted to depth orders, as the following lemma shows.

Lemma 10.4 *An order on S is a depth order in direction $\vec{d}$ for S if and only if the corresponding order on S° is a depth order in direction $\vec{d}$ for S°.*

Proof: Toussaint [119] has proved that a polygon $\mathcal{P}_i$ obscures a polygon $\mathcal{P}_j$ if and only if $\mathcal{CH}(\mathcal{P}_i|\mathcal{P}_j)$ obscures $\mathcal{CH}(\mathcal{P}_j|\mathcal{P}_i)$. By definition of relative convex hulls and by Lemma 10.1 (iii), we have $\mathcal{P}_i \subseteq \mathcal{P}_i^\circ \subseteq \mathcal{CH}(\mathcal{P}_i|\mathcal{P}_j)$ and $\mathcal{P}_j \subseteq \mathcal{P}_j^\circ \subseteq \mathcal{CH}(\mathcal{P}_j|\mathcal{P}_i)$. Hence, $\mathcal{P}_i$ obscures $\mathcal{P}_j$ if and only if $\mathcal{P}_i^\circ$ obscures $\mathcal{P}_j^\circ$. $\square$

10.2   Embeddings

Let $S = \{\mathcal{P}_1, \ldots, \mathcal{P}_m\}$ be a set of non-overlapping polygons in the plane. An *embedding* of S is a set $\tilde{S} = \{\tilde{\mathcal{P}}_1, \ldots, \tilde{\mathcal{P}}_m\}$ of non-overlapping polygons such that $\mathcal{P}_i \subseteq \tilde{\mathcal{P}}_i$, for $1 \leqslant i \leqslant m$. In this section we show that any set S can be embedded into a set $\tilde{S}$ that has few vertices and, moreover, that can be ordered if and only if S can be ordered. This embedding will help us to devise a storage-efficient data structure for depth order queries.

10.2.1   Embedding Convex Polygons

Let us start by considering a set $S = \{\mathcal{P}_1, \ldots, \mathcal{P}_m\}$ of disjoint (thus, they are not allowed to touch)[1] *convex* polygons, with n vertices in total. It is known that S can be embedded into a set $\tilde{S}$ with only $O(m)$ vertices in total. We sketch an algorithm due to Wenger [121] that computes such an embedding.

Augment S with three dummy triangles $\mathcal{P}_{m+1}$, $\mathcal{P}_{m+2}$ and $\mathcal{P}_{m+3}$ such that the convex hull of S consists of one vertex from each of $\mathcal{P}_{m+1}$, $\mathcal{P}_{m+2}$ and $\mathcal{P}_{m+3}$. Next we define a *triangulation* of S. Intuitively, this is a plane graph, where the nodes are the polygons and the arcs are segments between vertices of the polygons. More formally, a triangulation of a set S of polygons is a planar subdivision, consisting of the set S and some additional line segments—called *triangulation segments*—between vertices of polygons in S, such that each face which is not a polygon is bounded by exactly three triangulation segments and portions of the boundary of at most three polygons. If S consists of convex polygons, then the triangulation segments are necessarily between distinct polygons. See Figure 10.3 for an example. Let $\mathcal{G}$ be the planar graph whose nodes are the $m + 3$ polygons in S and whose edges are the triangulation segments. Since $\mathcal{G}$ is planar, the total number of edges in $\mathcal{G}$ is $O(m)$. (This is true even though there can be multiple edges between two nodes, because such edges are 'topologically' different.) For each pair of polygons $\mathcal{P}_i, \mathcal{P}_j$, we compute a line $l_{i,j}$, which separates $\mathcal{P}_i$ from $\mathcal{P}_j$. Let $L_1(\mathcal{P}_i) = \{l_{i,j} : \mathcal{P}_j \text{ is a neighbor of } \mathcal{P}_i \text{ in } \mathcal{G}\}$ be the set of lines separating $\mathcal{P}_i$ from its neighbors. Furthermore, let $l(e)$ be the line containing edge e of $\mathcal{G}$ and let $L_2(\mathcal{P}_i) = \{l(e) : e \text{ is a triangulation edge connecting two neighbors of } \mathcal{P}_i, e \text{ and } \mathcal{P}_i \text{ are on the boundary of a common face, and } l(e) \text{ does not intersect}[2] \mathcal{P}_i\}$. For each $1 \leqslant i \leqslant m$, let $\tilde{\mathcal{P}}_i$ be the intersection of the half-planes containing $\mathcal{P}_i$ that are bounded by lines in $L_1(\mathcal{P}_i)$ and $L_2(\mathcal{P}_i)$. Now $\tilde{S} = \{\tilde{\mathcal{P}}_1, \ldots, \tilde{\mathcal{P}}_m\}$. See Figure 10.3.

Lemma 10.5 *It is possible to compute an embedding $\tilde{S}$ of S with $O(m)$ vertices in time $O(n + m \log n)$.*

Proof: The fact that $\tilde{S}$ is an embedding with $O(m)$ vertices is not hard to prove. See [121]. It remains to show that $\tilde{S}$ can be computed in $O(n + m \log n)$ time.

First, we compute a triangulation of S. To this end we triangulate (in the usual sense) $\mathcal{CH}(S) - S$, the region in between the polygons, in $O(n + m \log n)$ time [12].

---

[1]This is not necessary, but it simplifies the description somewhat.
[2]The necessity of this condition was overlooked by Wenger [121].

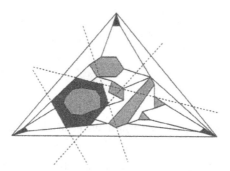

Figure 10.3: A triangulation of a set of polygons and the embedding for one of them. The black triangles are the dummy triangles.

Note that each face which is not a polygon contains either 2 or 3 triangulation segments on its boundary. To obtain a triangulation of S, we just remove triangulation segments that are not on the convex hull, until each face is bounded by three triangulation segments. Thus, we remove a segment if it bounds a face having only two triangulation segments on its boundary. Note that the resulting face has either 2 or 3 triangulation segments. Hence, when no more segments can be removed, we have obtained a triangulation of S. Since there are $O(n)$ segments in the triangulation of $\mathcal{CH}(S) - S$, it is straightforward to compute the triangulation of S in $O(n)$ time. The next step is to compute for each $\mathcal{P}_i$ the separating lines in $L_1(\mathcal{P}_i)$. One separating line $l_{i,j}$ can be computed in time $O(\log |\mathcal{P}_i| + \log |\mathcal{P}_j|) = O(\log n)$ by an algorithm of Edelsbrunner [49] or Chin and Wang [33]. The total time to compute all $O(m)$ separating lines is thus $O(m \log n)$. The intersection of the half-planes bounded by lines in $L_1(\mathcal{P}_i) \cup L_2(\mathcal{P}_i)$ can be computed in $O((|L_1(\mathcal{P}_i)| + |L_2(\mathcal{P}_i)|) \log(|L_1(\mathcal{P}_i)| + |L_2(\mathcal{P}_i)|))$ time [104], which adds up to $O(m \log m)$ time for all $\mathcal{P}_i$'s in total. □

Note that $\tilde{S}$ consists of convex polygons. Hence, it can still be ordered in any direction. Moreover, any depth order for $\tilde{S}$ is also valid for S (but not vice versa), since the polygons of S are contained in those of $\tilde{S}$.

10.2.2 Embedding Arbitrary Polygons

Let $S = \{\mathcal{P}_1, \ldots, \mathcal{P}_m\}$ be a set of non-overlapping simple polygons with n vertices in total. Our goal is to find an embedding of S with few vertices, that can be ordered if S can be ordered. In the convex case it is always possible to find an embedding with $O(m)$ vertices in total. In general this is not always possible. See Figure 10.4. However, we are able to compute an embedding that is good enough for our purposes. First we compute the set S° of relative convex hulls, using Algorithm 10.2. This way we make the polygons 'as convex as possible', which helps us to apply the same ideas that were used in the convex case. Two polygons in S° may share a number of

Figure 10.4: Two polygons for which any embedding has $\Omega(n)$ vertices.

edges as, for example, the relative convex hulls of the two polygons in Figure 10.4. Let E denote the set of edges that are shared by pairs of polygons in S°, where each edge is counted twice (once for each polygon that contains it). We will show how to construct an embedding $\tilde{S} = \{\tilde{\mathcal{P}}_1, \ldots, \tilde{\mathcal{P}}_m\}$ for S°, and thus for S, such that $\sum_{i=1}^{m} |\tilde{\mathcal{P}}_i| - |E| = O(m)$.

As in the convex case, we augment S° with three dummy triangles such that the convex hull of S° consists of one vertex of each of these triangles. Next, we triangulate the region $\mathcal{CH}(S^\circ) - S^\circ$. We want to remove certain segments from the triangulation of $\mathcal{CH}(S) - S$ to obtain a triangulation of S°. Recall that in a triangulation of a set of polygons it is required that each face is bounded by three triangulation segments. To meet this requirement when the polygons are allowed to touch, we have to add degenerate triangulation segments between touching polygons. More specifically, for each maximal chain of boundary edges that is shared by two polygons we add the first and last vertex of this chain as degenerate triangulation segments. Two degenerate triangulation segments that coincide—which happens when two polygons share only one vertex—are considered to be different. Now the same procedure as in the convex case can be used to obtain a triangulation of S°: remove non-degenerate triangulation segments that bound a face with only two triangulation segments on its boundary (and which are not on the convex hull), until each face is bounded by exactly three triangulation segments. The result is a set of $O(m)$ triangulation segments. Note that there may be triangulation segments between vertices of the same polygon.

The boundary edges of a polygon in S° that are not shared with other polygons in S° form a number of convex chains. Let γ be such a chain. Note that γ does not share vertices with other polygons, except possibly the first and last vertex. Chain γ is on the boundary of a number of faces of the triangulation, and a non-degenerate triangulation segment t that is incident to γ separates two such faces.

Next we construct a chain $\tilde{\gamma}$ with few vertices that replaces γ in the embedding. We assume that the first and last vertex of γ are shared with other polygons; the adaptation to the simpler case where γ does not share a vertex with any other polygon is straightforward. Let $f_1(\gamma), \ldots, f_a(\gamma)$ be an enumeration in clockwise order of the faces that are bounded by γ, and define $sleeve(\gamma)$ to be the concatenation of these faces, as in Section 10.1. Recall that $f_1(\gamma)$ and $f_a(\gamma)$ are not concatenated, even when they share an edge. Thus we can consider $sleeve(\gamma)$ to be a simple polygon. Furthermore, let $t_i(\gamma)$, $1 \leqslant i \leqslant a - 1$, be the triangulation segment separating $f_i(\gamma)$ from $f_{i+1}(\gamma)$,

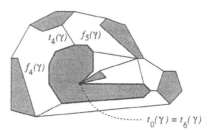

Figure 10.5: The sleeve of the fat chain γ.

and let $t_0(\gamma)$ and $t_a(\gamma)$ be the degenerate triangulation segments incident to the first and last vertex of γ. See Figure 10.5 for an illustration of these definitions. Note that $t_0(\gamma)$ coincides with $t_a(\gamma)$ in the example of Figure 10.5. A triangulation segment $t_i(\gamma)$ connects γ to some other chain γ'. We will construct a separator for $t_i(\gamma)$, which is a segment that lies inside $sleeve(\gamma)$ and separates γ from γ' inside $sleeve(\gamma)$. The chain $\tilde{\gamma}$ is constructed from these separators, and from some of the triangulation segments between neighbors of γ. Separators are defined as follows. Consider a line segment $s \subset sleeve(\gamma)$ that touches the boundary of $sleeve(\gamma)$ with both endpoints. Such a segment s partitions $sleeve(\gamma)$ into two regions. We call s a *separator* for $t_i(\gamma)$ if γ and γ' are contained in different regions. More precisely, we require the parts of γ and γ' that are on the boundary of the two faces $f_i(\gamma)$ and $f_{i+1}(\gamma)$ to be in different regions. We call these parts the *relevant parts* of γ and γ' with respect to $t_i(\gamma)$. We denote the separator for $t_i(\gamma)$ by $sep(t_i(\gamma))$.

Lemma 10.6 *For every triangulation segment $t_i(\gamma)$, a separator $sep(t_i(\gamma))$ exists. Moreover, all the separators $sep(t_0(\gamma)), \ldots, sep(t_a(\gamma))$ for the triangulation segments $t_0(\gamma), \ldots, t_a(\gamma)$ incident to γ can be computed in $O(sleeve(|\gamma|) \log |sleeve(\gamma)|)$ time in total.*

Proof: Consider a non-degenerate triangulation segment $t_i(\gamma)$ in $sleeve(\gamma)$. Let s be a segment of maximal length inside $sleeve(\gamma)$ that is tangent to γ at the endpoint of $t_i(\gamma)$. This segment partitions $sleeve(\gamma)$ into two regions, and the relevant part of γ is fully contained in one of them. We call this region the *inner region* of s, and the other region the *outer region*. Now, if s does not have an endpoint on γ', then the relevant part of γ' is fully contained in the outer region, and we let $sep(t_i(\gamma)) = s$. Otherwise, let q' be the endpoint of s that is on γ'. Because γ' is a convex chain, there is only one such endpoint. Denote the point where s touches γ by q. When we rotate s, such that s remains tangent to γ, then q moves along γ, and q' moves along γ'. There are two directions into which we can move q'; we choose the direction of rotation such that the portion of γ' in the outer region grows. Clearly, the point q' must leave $f_i(\gamma)$ some time, before q reaches the point where $t_{i-1}(\gamma)$ touches γ. Thus q' starts on γ', and at the end it is not on (the relevant part of) γ'. Hence, at some point during the rotation, s must be tangent to γ'. When this happens, then (the relevant part of) γ'

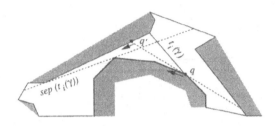

Figure 10.6: Construction of a separator.

is fully contained in the outer region and s separates γ from γ'. See Figure 10.6 for an illustration.

For degenerate triangulation segments $t_i(\gamma)$ —note that only $t_0(\gamma)$ and $t_a(\gamma)$ are possibly degenerate—we construct $sep(t_i(\gamma))$ as follows. Without loss of generality, consider $t_0(\gamma)$, and suppose that one of the two polygons involved has a reflex vertex that is incident to $t_0(\gamma)$; if none of the polygons has a reflex vertex, then we can use the same construction as in the non-degenerate case. To obtain $sep(t_0(\gamma))$, we extend the edge of this polygon that is incident to $t_0(\gamma)$ and bounds $f_1(\gamma)$, until it hits the boundary of $sleeve(\gamma)$. See Figure 10.7. Notice that the angle of the reflex vertex does

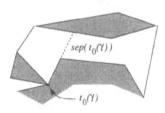

Figure 10.7: Construction of a separator for a degenerate triangulation segment.

not change. This is important for our application, because changing the angle of a reflex vertex might introduce a cycle for a certain direction that was cycle-free before.

We have proved the existence of separators. Now let us see how to construct them efficiently. To this end, we preprocess $sleeve(\gamma)$ for $O(\log|sleeve(\gamma)|)$ time ray shooting queries in $O(|sleeve(\gamma)|\log|sleeve(\gamma)|)$ time [27]. Consider the computation of separator $sep(t_i(\gamma))$ for a non-degenerate triangulation segment $t_i(\gamma)$. By ray shooting we can find in $O(\log|sleeve(\gamma)|)$ time the points where a segment that touches γ hits the boundary of $sleeve(\gamma)$. The rotation of segment s until it becomes tangent to γ' is easily implemented in time that is linear in the number of vertices of the relevant parts of γ and γ'. The separator for a degenerate triangulation segment can be found in $O(\log|sleeve(\gamma)|)$ time by ray shooting. Summing over all triangulation segments $t_i(\gamma)$ that are incident to γ, we obtain $O(|sleeve(\gamma)|\log|sleeve(\gamma)|)$ time in total. $\square$

Now we are ready to construct $\tilde{\gamma}$. The construction is done with the following incremental algorithm. At the start of the algorithm $\tilde{\gamma}$ consists of $sep(t_0(\gamma))$. The remaining separators $sep(t_1(\gamma)), \ldots, sep(t_a(\gamma))$ are processed in order; meanwhile we update $\tilde{\gamma}$. Suppose $sep(t_0(\gamma)), \ldots, sep(t_{i-1}(\gamma))$ already have been added. Recall that $sep(t_i(\gamma))$ partitions $sleeve(\gamma)$ into two regions. To process $sep(t_i(\gamma))$, we have to remove the part of $\tilde{\gamma}$ that is on the wrong side of $sep(t_i(\gamma))$—that is, the part that lies in the outer region—and replace this part by the relevant portion of $sep(t_i(\gamma))$. The part of $\tilde{\gamma}$ that has to be removed can be found by doing a binary search with the slope of $sep(t_i(\gamma))$ in the slopes of the segments forming $\tilde{\gamma}$. If the relevant portion of $sep(t_i(\gamma))$ intersects a triangulation segment between neighbors of γ (this intersection is necessarily an endpoint of $sep(t_i(\gamma))$), then add the relevant portion of this triangulation segment as well. The result for the chain of Figure 10.5 can be seen in Figure 10.8. To construct the embedding $\tilde{S}$, we replace every chain γ on the boundary

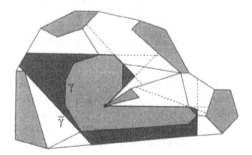

Figure 10.8: The chain $\tilde{\gamma}$ that replaces γ, and the resulting embedding for the polygon.

of each relative convex hull by the chain $\tilde{\gamma}$. For each triangulation segment t we only construct one separator, which is used in the construction of the replacements of both chains incident to t. In other words, if $sep(t)$ is the separator for γ_t and γ_t' that is created in the construction of $\tilde{\gamma}$, then $sep(t)$ is also used in the construction of $\tilde{\gamma}_t'$. Note, however, that we cannot use $sep(t)$ directly, since $sleeve(\gamma)$ and $sleeve(\gamma')$ are different. But using the ray shooting structure for $sleeve(\gamma')$ it is easy to compute the 'new' separator. Thus the algorithm to construct $\tilde{S}$ is as follows.

Algorithm 10.7
Input: A set S of non-overlapping polygons.
Output: An embedding $\tilde{S}$ of S with few vertices.
1. Compute the set S° of relative convex hulls of the polygons in S.
2. Compute a triangulation of S°.
3. **for** every convex chain γ of each polygon $\mathcal{P}_i^\circ \in S^\circ$
4.   **do** Preprocess $sleeve(\gamma)$ for efficient ray shooting.
5.   **for** every triangulation segment $t_i(\gamma)$ that is incident to γ
6.   **do if** a separator $sep(t_i(\gamma))$ for $t_i(\gamma)$ has not been computed before
7.   **then** Compute $sep(t_i(\gamma))$ as described above.
8.   **else** Compute $sep(t_i(\gamma))$ from the old separator.
9. $\tilde{\gamma} \leftarrow sep(t_0(\gamma))$
10.   **for** $i = 1$ **to** a
11.   **do** Remove the part of $\tilde{\gamma}$ that is on the wrong side of $sep(t_i(\gamma))$.
12.   Add the relevant portion of $sep(t_i(\gamma))$ to $\tilde{\gamma}$.
13.   **if** $sep(t_i(\gamma))$ intersects a triangulation segment t
14.   **then** add the relevant part of t to $\tilde{\gamma}$.
15. Replace every chain γ by $\tilde{\gamma}$.

The following lemma is important for the correctness of our algorithm. Recall that a is the number of triangulation segments incident to γ.

Lemma 10.8 $\tilde{\gamma}$ *is convex chain with* $O(a)$ *vertices that can be computed in time* $O(|sleeve(\gamma)| \log |sleeve(\gamma)|)$.

Proof: The $O(a)$ bound on the size of $\tilde{\gamma}$ immediately follows from the construction. We show that the separator $sep(t_i(\gamma))$ that is processed always intersects the current chain $\tilde{\gamma}$. To this end, consider $sep(t_{i-1}(\gamma))$. We claim that $sep(t_i(\gamma))$ either intersects $sep(t_{i-1}(\gamma))$, or that $sep(t_i(\gamma))$ and $sep(t_{i-1}(\gamma))$ intersect the same triangulation segment between two neighbors of γ. It then follows that the current chain $\tilde{\gamma}$ is intersected by $sep(t_i(\gamma))$. To prove our claim, consider the separators $sep(t_i(\gamma))$ and $sep(t_{i-1}(\gamma))$. We know that $sep(t_{i-1}(\gamma))$ separates the relevant parts of γ and some chain γ', that is, the parts inside $f_{i-1}(\gamma) \cup f_i(\gamma)$. Also, $sep(t_i(\gamma))$ separates γ from some chain γ'' inside $f_i(\gamma) \cup f_{i+1}(\gamma)$. This means that either $sep(t_i(\gamma))$ intersects $sep(t_{i-1}(\gamma))$, or they both intersect the triangulation segment connecting γ' to γ''. Thus $sep(t_i(\gamma))$ intersects the current chain, from which it follows that $\tilde{\gamma}$ is one chain. We leave it to the reader to verify that $\tilde{\gamma}$ is a *convex* chain.

To prove the construction time, we note that a separator $sep(t_i(\gamma))$ is processed in $O(\log |\tilde{\gamma}| + k_i)$ time, where k_i is the number of separators that are removed from the current chain in line 11 of the algorithm. A separator will not return when it has been removed, so $\sum_{i=1}^{a} k_i \leqslant a$. The time bound now follows from Lemma 10.6. $\qquad\square$

Theorem 10.9 *Let S be a set of m simple non-overlapping polygons with n vertices in total, and E be the set of edges shared by the relative convex hulls of the polygons in S, where each edge is counted twice (once for each polygon that contains it). An* embedding $\tilde{S} = \{\tilde{\mathcal{P}}_1, \ldots, \tilde{\mathcal{P}}_m\}$ of S with $\left(\sum_{1 \leqslant i \leqslant m} |\tilde{\mathcal{P}}_i|\right) - |E| = O(m)$ can be computed in $O(n \log n)$ time.

Proof: The time bound follows from Lemma 10.8 and the fact that $\sum_{\gamma} |sleeve(\gamma)| = O(n)$. The bound on the size of $\tilde{S}$ follows from Lemma 10.8 and the fact that the total number of triangulation segments is $O(m)$. It remains to prove that $\tilde{S}$ is indeed an embedding.

It is immediate that each polygon $\mathcal{P}_i \in S$ is contained in $\tilde{\mathcal{P}}_i$. To complete the proof we must show that the polygons in $\tilde{S}$ do not intersect each other. Consider a polygon $\tilde{\mathcal{P}}_i$. Since boundary chains are only replaced by chains inside the corresponding sleeves, $\tilde{\mathcal{P}}_i$ can only intersect another polygon $\tilde{\mathcal{P}}_j$ somewhere inside a face that belongs to the sleeves of both $\tilde{\mathcal{P}}_i$ and $\tilde{\mathcal{P}}_j$. But this cannot happen because $\tilde{\mathcal{P}}_i$ is separated from $\tilde{\mathcal{P}}_j$ by the separator that is constructed for the triangulation segment between $\tilde{\mathcal{P}}_i$ and $\tilde{\mathcal{P}}_j$. $\square$

We have seen that we can embed a set of arbitrary polygons into another set of polygons with few vertices. But we also want to be able to order the new set. To prove that this is possible—if it is possible for the original set of polygons—we need the following lemma, proved by Toussaint [118].

Lemma 10.10 (Toussaint [118]) *A depth order for a set of polygons exists if and only if there exists a depth order for every pair of polygons in the set.*

Now we can prove that the embedding can still be ordered.

Lemma 10.11 $\tilde{S}$ *can be ordered in direction* $\vec{d}$ *if and only if* S *can be ordered in direction* $\vec{d}$, *and any depth order for* $\tilde{S}$ *is also valid for* S.

Proof: Since $\mathcal{P}_i \subseteq \tilde{\mathcal{P}}_i$ for every $\mathcal{P}_i \in S$, the only non-trivial part of the lemma is that $\tilde{S}$ can be ordered if S can be ordered. By Lemma 10.4, it suffices to show that $\tilde{S}$ can be ordered if S° can be ordered. Suppose for a contradiction that S° can be ordered in direction $\vec{d}$ but $\tilde{S}$ cannot be ordered. By Lemma 10.10, there are two polygons $\tilde{\mathcal{P}}_i$ and $\tilde{\mathcal{P}}_j$ that cannot be ordered. Assume without loss of generality that $\vec{d}$ is vertically upward. Thus, $\tilde{\mathcal{P}}_j$ collides with $\tilde{\mathcal{P}}_i$ when it is moved upward, and also when it is moved downward. Let a and b be the points on the boundary of $\tilde{\mathcal{P}}_i$ where $\tilde{\mathcal{P}}_j$ collides with $\tilde{\mathcal{P}}_i$ when it is moved upward and downward, respectively. Let l_a and l_b be the vertical lines through these two points. The points a and b are both on $\partial\tilde{\mathcal{P}}_i$. Hence, $\partial\tilde{\mathcal{P}}_i$ either crosses l_a in a point a' below $\tilde{\mathcal{P}}_j$ or it crosses l_b in a point b' above $\tilde{\mathcal{P}}_j$. Assume without loss of generality that the first case occurs. See Figure 10.9. The part of the boundary of $\tilde{\mathcal{P}}_i$ connecting a and a' and not containing b must contain a reflex vertex v such that both edges incident to v lie on one side of the vertical line through v. But any reflex vertex of $\tilde{\mathcal{P}}_i$ is a reflex vertex of $\mathcal{P}_i^\circ$ as well and, moreover, the angles at these vertices are equal. By Lemma 10.1(ii), this is a vertex of some other polygon $\mathcal{P}_k^\circ$ or a vertex of $\mathcal{P}_i^\circ$ that is used twice. In the first case, $\mathcal{P}_i^\circ$ and $\mathcal{P}_k^\circ$ cannot be ordered, contradicting the fact that S° can be ordered. We leave it to the reader to verify that the second case leads to a contradiction as well. $\square$

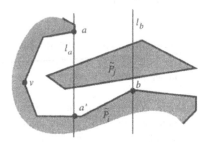

Figure 10.9: Illustration of the proof of Lemma 10.11.

10.3 Computing All Cycle-Free Directions

In this section it is shown that all cycle-free directions—that is, all directions for which a depth order exists—can be computed in $O(n \log n)$ time. Toussaint [119] has shown that there exists a depth order for two polygons in direction $\vec{d}$ if and only if their relative convex hulls are monotone in direction $\vec{d} + \frac{1}{2}\pi$. He uses this result to compute all cycle-free directions of two polygons. Lemma 10.10 implies that the fact stated above for two polygons also holds for larger sets of polygons:

Lemma 10.12 *There exists a depth order in direction $\vec{d}$ for a set of polygons if and only if the relative convex hulls of the polygons are monotone in direction $\vec{d} + \frac{1}{2}\pi$.*

The proof is not difficult and therefore omitted. Monotonicity of a polygon can be characterized as follows:

Observation 10.13 *A polygon is monotone in direction $\vec{d} + \frac{1}{2}\pi$ if and only if it has no reflex vertex v such that the two edges incident to v lie on the same side of the line through v with slope $\vec{d}$.*

For a reflex vertex v of some $\mathcal{P}^\circ \in S^\circ$, let $I_v \subset [0 : 2\pi]$ be the interval such that the two edges incident to v lie on the same side of a line through v with slope $\vec{d}$ if and only if $\vec{d} \in I_v$. (In fact, I_v can consist of two disjoint intervals, one starting at 0, the other ending at 2π.) Given the two edges incident to v, I_v is easily computed in constant time. Thus, we can compute $I(S^\circ) = \{I_v | v \text{ is a reflex vertex of a } \mathcal{P}^\circ \in S^\circ\}$ in $O(n)$ time. By Lemma 10.12 and Observation 10.13, a depth order for S in direction $\vec{d}$ exists if and only if $\vec{d} \notin \bigcup I(S^\circ)$. In other words, the set D of directions for which a depth order exists is the set $[0 : 2\pi] - \bigcup I(S^\circ)$. This leads to:

Theorem 10.14 *All directions for which a depth order exists for a given set S of polygons with a total number of n vertices can be determined in time $O(n \log n)$. Furthermore, there exists a data structure that uses $O(n)$ storage, such that we can decide in $O(\log n)$ time whether a depth order exists in a query direction.*

Proof: S° can be computed in time $O(n \log n)$ by Theorem 10.3. After that we can compute $I(S^\circ)$ in linear time. By sorting the endpoints of the intervals in $I(S^\circ)$ and performing a line sweep, keeping track of the number of intervals currently intersected by the sweep point, the set $D = [0 : 2\pi] - \bigcup(S^\circ)$ can be found in time $O(n \log n)$. The data structure for deciding whether a depth order exists in a query direction is a balanced binary search tree on the endpoints of the intervals that form D. Clearly, this structure has $O(\log n)$ query time and it uses $O(n)$ storage. □

10.4 Depth Order Queries in the Plane

Recall that a depth order in direction $\vec{d}$ for a set S of polygons is an order such that no collisions occur if the polygons are moved one at a time in direction $\vec{d}$ to infinity according to this order. Thus, the computation of depth orders involves computing some sort of dominance relation between the polygons, where a polygon $\mathcal{P}$ dominates another polygon $\mathcal{P}'$ if $\mathcal{P}'$ collides with $\mathcal{P}$ when it is moved into direction $\vec{d}$. A depth order exists if and only if the dominance relation between the polygons is free of cycles. It has been shown by Guibas and Yao [65] that it is not necessary to compute all (possibly $\Omega(m^2)$) dominances explicitly, but that it suffices to compute the *immediate* dominances: a polygon $\mathcal{P}$ immediately dominates $\mathcal{P}'$ if, when $\mathcal{P}'$ is moved, it intersects $\mathcal{P}$ before it intersects any other polygon. This immediate dominance relation changes radically, however, when the direction $\vec{d}$ changes. Hence, if we want to do preprocessing to speed up the computation of a depth order for a query direction $\vec{d}$, we have to take a different approach. The basic idea is that a triangulation of the area in between the polygons gives us all the information we need to compute a depth order for any given direction. Furthermore, instead of translating the set of polygons itself, we compute an embedding and translate the polygons in the embedding. This reduces the amount of storage and speeds up the queries.

10.4.1 Convex Polygons

Let $S = \{\mathcal{P}_1, \ldots, \mathcal{P}_m\}$ be a set of m non-overlapping convex polygons with n vertices in total. First, we compute an embedding $\tilde{S}$, according to Lemma 10.5, and we replace S by $\tilde{S}$. Observe that $\tilde{S}$ consists of convex polygons, so $\tilde{S}$ can still be translated into every direction [65]. Moreover, since the polygons of S are contained in those of $\tilde{S}$, any depth order for $\tilde{S}$ is also valid for S.

Let $T = \{t_1, \ldots, t_k\}$ be the set of triangles in a triangulation of $\mathcal{CH}(\tilde{S}) - \tilde{S}$, the area in between the polygons of $\tilde{S}$. The idea is to translate the set $\tilde{S} \cup T$. Observe that this new set again only contains convex polygons and, hence, it can still be translated. Surprisingly, translating $\tilde{S} \cup T$ is an easier task than translating $\tilde{S}$, as follows from Lemma 10.15 below. Let us define $\vec{d}$-neighbors, a concept that is crucial in our method. Let $\mathcal{P}$ and $\mathcal{P}'$ be two polygons. $\mathcal{P}$ is a $\vec{d}$-*neighbor* of $\mathcal{P}'$ if and only if (i) there are edges e of $\mathcal{P}$ and e' of $\mathcal{P}'$ such that $e \cap e' \neq \varnothing$, and (ii) there is a ray in direction $\vec{d}$ that intersects $int(\mathcal{P}')$ just before it intersects $e \cap e'$ and $int(\mathcal{P})$ just after it intersects $e \cap e'$. Notice that if two polygons $\mathcal{P}$ and $\mathcal{P}'$ share an edge e, then either $\mathcal{P}$ is a $\vec{d}$-neighbor of $\mathcal{P}'$, or $\mathcal{P}'$ is a $\vec{d}$-neighbor of $\mathcal{P}$, or e is parallel to $\vec{d}$. See Figure 10.10 for an illustration of this definition.

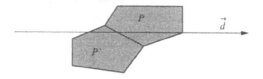

Figure 10.10: $\mathcal{P}$ is a $\vec{d}$-neighbor of $\mathcal{P}'$.

Lemma 10.15 *A polygon (possibly a triangle) $\mathcal{P} \in \tilde{S} \cup T$ can be translated to infinity in direction $\vec{d}$ without collisions if and only if all its $\vec{d}$-neighbors already have been translated without collisions.*

Proof: The 'only if'-part is trivial. To prove the 'if'-part, suppose that all $\vec{d}$-neighbors of $\mathcal{P}$ have been translated without collisions, but that $\mathcal{P}$ still collides with some polygon $\mathcal{P}'$. Consider the moment that $\mathcal{P}$ and $\mathcal{P}'$ first intersect during the translation. This intersection involves an edge e of $\mathcal{P}$. But then the $\vec{d}$-neighbor of $\mathcal{P}$ that shares e—which must exist since the area in between $\mathcal{P}$ and $\mathcal{P}'$ has been triangulated—also collides with $\mathcal{P}$, which contradicts the assumptions. $\square$

Lemma 10.15 immediately leads to the following simple scheme. The preprocessing just consists of computing a triangulation T of $\mathcal{CH}(\tilde{S}) - \tilde{S}$ and the dual graph $\mathcal{G} = \mathcal{G}(\tilde{S} \cup T)$ of $\tilde{S} \cup T$. The nodes in this graph correspond to the polygons in $\tilde{S} \cup T$ and there is an arc between two nodes if the corresponding polygons share an edge. Given a query direction $\vec{d}$, we proceed as follows. First, we turn $\mathcal{G}$ into a directed graph $\mathcal{G}_{\vec{d}}$. Let a be an arc in $\mathcal{G}$ connecting nodes corresponding to polygons $\mathcal{P}$ and $\mathcal{P}'$. If $\mathcal{P}$ is a $\vec{d}$-neighbor of $\mathcal{P}'$ then the arc in $\mathcal{G}_{\vec{d}}$ corresponding to a, denoted $a_{\vec{d}}$, is directed from $\mathcal{P}'$ to $\mathcal{P}$. If $\mathcal{P}'$ is a $\vec{d}$-neighbor of $\mathcal{P}$ then $a_{\vec{d}}$ is directed from $\mathcal{P}$ to $\mathcal{P}'$. Otherwise the edge shared by $\mathcal{P}$ and $\mathcal{P}'$ is parallel to $\vec{d}$, and a has no corresponding arc in $\mathcal{G}_{\vec{d}}$. Thus a node corresponding to some polygon has outgoing arcs to all its $\vec{d}$-neighbors and incoming arcs from all polygons for which it is a $\vec{d}$-neighbor. From Lemma 10.15 and the definition of $\mathcal{G}_{\vec{d}}$ it easily follows that a topological order of the nodes in $\mathcal{G}_{\vec{d}}$ corresponds to a depth order in direction $\vec{d}$ for the polygons in $\tilde{S} \cup T$. Note that the fact that $\tilde{S} \cup T$ can be ordered guarantees that $\mathcal{G}_{\vec{d}}$ is acyclic. Clearly, if the triangles of T are omitted from of this order we get a depth order for $\tilde{S}$, which corresponds to an order for S. This leads to:

Theorem 10.16 *A set of m convex polygons with n vertices in total can be preprocessed in $O(n + m \log n)$ time into a data structure of size $O(m)$ such that, given a direction $\vec{d}$, a depth order for S in direction $\vec{d}$ can be computed in time $O(m)$.*

Proof: The embedding $\tilde{S}$ of S can be computed in $O(n + m \log n)$ time, according to Lemma 10.5. The convex hull of $\tilde{S}$ as well as the triangulation and its dual graph can be computed in time $O(m \log m)$ [12, 104]. Note that the total number of edges in $\tilde{S} \cup T$, and therefore the number of nodes and arcs in $\mathcal{G}$, is $O(m)$. Since we can decide in constant time what the direction of an arc $a_{\vec{d}}$ in $\mathcal{G}_{\vec{d}}$ is, the construction of $\mathcal{G}_{\vec{d}}$ takes only linear time. Topologically sorting a directed acyclic graph can also be done in linear time. See, for example, Knuth [74]. $\qquad \Box$

10.4.2 Arbitrary Polygons

We will now show how to compute depth orders for a set S of arbitrary polygons. Again, we replace S by its embedding $\tilde{S}$, according to Theorem 10.9. Note that the idea of triangulating the area in between the polygons will not work with an arbitrary set of polygons. The problem is that if there are non-convex polygons, the triangles of

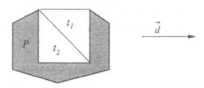

Figure 10.11: $\{\mathcal{P}\}$ can be ordered, but $\{\mathcal{P}, t_1, t_2\}$ cannot be ordered.

the triangulation might prevent the existence of a depth order. That is, it is possible that a depth order exists for S, but not for $S \cup T$. See Figure 10.11. Fortunately, this problem does not arise for the set $\tilde{S}$. Let T be a triangulation of $\mathcal{CH}(\tilde{S}) - \tilde{S}$.

Lemma 10.17 *There exists a depth order in direction $\vec{d}$ for $\tilde{S}$ if and only if there exists a depth order in direction $\vec{d}$ for $\tilde{S} \cup T$.*

Proof: The 'if'-part is trivial. The proof of the 'only if'-part is similar to the proof of Lemma 10.11. $\qquad \Box$

We thus arrive at the following scheme for ordering a set S of arbitrary polygons. As a first preprocessing step, we compute an embedding $\tilde{S}$ of S with few vertices according to Theorem 10.9. Next, we compute a triangulation T of $\mathcal{CH}(\tilde{S}) - \tilde{S}$, together with its dual graph $\mathcal{G} = \mathcal{G}(\tilde{S} \cup T)$. Note that the number of arcs in $\mathcal{G}$ is linear in the size of $\tilde{S} \cup T$, which is $O(m)$ by Theorem 10.9. Furthermore, we compute for each arc a in $\mathcal{G}$ some information that will help us to direct a for a query direction. Recall that a corresponds to two polygons $\mathcal{P}, \mathcal{P}'$ that share an edge or a chain of edges. Each edge e on the shared chain induces an interval $I_e = [\theta_e : \theta_e + \pi]$ such that $\mathcal{P}$ collides with $\mathcal{P}'$ along e when moved into a direction that lies in I_e, and $\mathcal{P}'$ collides with $\mathcal{P}$ in the opposite directions. By taking the union of the intervals I_e we find the directions where $\mathcal{P}$ collides with $\mathcal{P}'$ along the chain. Similarly, by taking the union of the opposite intervals $[0 : 2\pi] - I_e$ we find the directions where $\mathcal{P}'$ collides with $\mathcal{P}$ along the chain. Clearly, the polygons are interlocked in directions that lie in the intersection of these unions. The remaining directions are partitioned into two opposite intervals, corresponding to directions where $\mathcal{P}$ can be moved without colliding with $\mathcal{P}'$ along the chain, and directions where $\mathcal{P}'$ can be moved without colliding with $\mathcal{P}$ along the chain. We store these intervals I_a and I'_a, which can be computed in linear time in the number of edges of the common chain, with the arc a. Given a query direction $\vec{d}$, we can construct $\mathcal{G}_{\vec{d}}$ in $O(m)$ time using the intervals I_a. If we find that two polygons are interlocked, then no depth order exists. Otherwise, we try to sort $\mathcal{G}_{\vec{d}}$ topologically. A topological order corresponds to a depth order for $\tilde{S}$ which, by Lemma 10.11, corresponds to a depth order for $\tilde{S}$. (Note that Lemma 10.15 is true for non-convex polygons too.) If $\mathcal{G}_{\vec{d}}$ cannot be sorted because it contains a cycle, then $\tilde{S} \cup T$ cannot be ordered. By Lemma's 10.11 and 10.17 we can then conclude that no depth order for S exists either. We conclude with the following theorem.

Theorem 10.18 *A set of m arbitrary polygons with n vertices in total can be pre-processed in $O(n \log n)$ time into a data structure of size $O(m)$ such that, given a direction $\vec{d}$, a depth order for S in direction $\vec{d}$ can be computed in time $O(m)$, if one exists.*

Remark 10.19 We can combine this results with Theorem 10.14 if we want to decide more quickly whether a depth order exists. Note, however, that this will increase the storage requirement of the structure to $O(n)$.

10.5 Depth Order Queries for Terrains

In this section we use the data structure of the previous section to generate depth orders with respect to a query point for sets of convex polygonal faces whose orthogonal projections onto the xy-plane do not intersect. Thus we can compute depth orders to generate *perspective* views for polygonal terrains consisting of convex faces using the painter's algorithm. In fact, we can handle more general scenes, since we do not require the scene to be connected as is the case for terrains. Let $F = \{f_1, \ldots, f_m\}$ be a set of convex polygonal faces in $\mathbb{E}^3$, with non-intersecting orthogonal projections onto the xy-plane. Let $\overline{F} = \{\overline{f}_1, \ldots, \overline{f}_m\}$ be the set of these projections. We will show that after computing an embedding $\tilde{F}$ of $\overline{F}$, a triangulation T of $\mathcal{CH}(\overline{F}) - \overline{F}$, and its dual graph $\mathcal{G} = \mathcal{G}(\tilde{F} \cup T)$, a depth order with respect to a query point p whose orthogonal projection $\overline{p}$ lies outside $\mathcal{CH}(\overline{F})$ can be computed in linear time.

To find a depth order for the faces of $\tilde{F}$ with respect to a query point, all that we have to change in the algorithms of the previous section is the concept of neighbors. Let $\tilde{f}$ and $\tilde{f}'$ be two polygons and $\overline{p}$ be a point in the plane. $\tilde{f}$ is a $\overline{p}$-*neighbor* of $\tilde{f}'$ if and only if (i) there are edge e of $\tilde{f}$ and e' of $\tilde{f}'$ such that $e \cap e' \neq \varnothing$, and (ii) there is a ray starting at $\overline{p}$ and intersecting $e \cap e'$ that intersects $int(\tilde{f}')$ just before intersecting $e \cap e'$ and $int(\tilde{f})$ just after intersecting $e \cap e'$. When we replace the notion of $\vec{d}$-neighbor in Lemma 10.15 by $\overline{p}$-neighbor, then Lemma 10.15 holds for depth orders with respect to point $\bar{p}$. This means that Theorem 10.16 also holds for depth order queries with respect to a query point that lies outside the convex hull of the set of polygons. The following lemma shows that the same approach can be used for terrains.

Lemma 10.20 *A depth order on $\tilde{F}$ with respect to $\bar{p}$ corresponds to a depth order on F with respect to p.*

Proof: Let $f_i, f_j \in F$ be two faces such that $f_i \prec_p f_j$. Thus, there is a ray ρ starting at p that first intersects f_i and then f_j. Obviously, $\bar{p}$ intersects both $\bar{f}_i$ and $\bar{f}_j$. Moreover, because $\bar{f}_i \cap \bar{f}_j = \varnothing$, the order of intersection does not change. Hence, $\bar{f}_i \prec_{\bar{p}} \bar{f}_j$, which implies $\tilde{f}_i \prec_{\bar{p}} \tilde{f}_j$. We conclude that a depth order on $\tilde{F}$ with respect to $\bar{p}$ corresponds to a depth order on F with respect to p. $\qquad\square$

In the following theorem, we assume that $\bar{p}$, the orthogonal projection of the query point p onto the xy-plane, lies outside $\mathcal{CH}(\tilde{F})$.

Theorem 10.21 *A terrain consisting of m convex polygonal faces with n vertices in total can be preprocessed in $O(n + m \log n)$ time into a data structure of size $O(m)$ such that a depth order on the faces of F with respect to a query point p can be determined in $O(m)$ time.*

Remark 10.22 If the terrain contains non-convex faces we can always cut these faces into convex parts. The restriction to convex faces is necessary for the following reason: if there are non-convex faces, then it is possible that a depth order exists for the faces of the terrain, but not for the corresponding two-dimensional problem. Consider, for example, the case where each face of the terrain is contained in the xy-plane and the faces are such that they cannot be ordered in the plane. In spite of this, a valid displaying order exists for view points above the terrain; in fact, any order is valid.

Remark 10.23 The assumption that $\bar{p}$, the orthogonal projection of p onto the xy-plane, does not lie inside the convex hull of $\tilde{F}$ ensures that a depth order for $\tilde{F}$ with respect to $\bar{p}$ always exists. This restriction can be removed if we are willing to decompose some of the faces of the terrain into smaller pieces. In particular, we can split $\tilde{F} \cup T$ into two sets with a line through $\bar{p}$. It is straightforward to compute the data structures for these two sets in $O(m)$ time from the data structure for the original set. Now a depth order can be found for the two sets separately. For the application to hidden surface removal using the painter's algorithm such an approach is satisfactory.

Chapter 11

Depth Orders in Three Dimensions

In this chapter we show how to compute or verify depth orders in three-dimensional space. The algorithms that we give are surprisingly simple. They are based on a general framework for computing a linear extension of a relation $(S, \prec)$. It is easy to compute an extension in time that is linear in the number of pairs that are related; to this end one topologically sorts the directed graph $\mathcal{G} = (S, E)$, where $(o, o') \in E$ if and only if $o \prec o'$. This is the approach taken in [21, 91] to sort a set S of n segments in space: first compute all pairs of segments that are related—which can be done in $O(n \log n + i)$ time by computing all i intersections in the projection plane—and then sort the corresponding graph $\mathcal{G}$ in $O(n + i)$ time. If $(S, \prec)$ does not contain a cycle then the sorting will succeed, otherwise some cycle will be detected in the graph $\mathcal{G}$. Note that i, the number of edges in $\mathcal{G}$, can be $\Theta(n^2)$. We show that it is not necessary to compute the full graph corresponding to $(S, \prec)$. All that is needed is to have a data structure that answers the following question: Given an element $o \in S$, return a predecessor of o and a successor of o, if they exist. The data structure should allow for the deletion of elements from S in sublinear time. In cases where the relation is given implicitly—such as for depth orders—this is often possible. Our algorithm uses an interesting form of divide-and-conquer, where the divide-step does not need to balanced. In fact, the more unbalanced it is, the better the running time of the algorithm.

11.1 Computing Linear Extensions

Let $\prec$ be a binary relation defined on a set S of n objects. Note that $\prec$ does not necessarily define a partial order on S, since we do not assume transitivity. This will be useful in our application, because the relations $\prec_p$ or $\prec_{\vec{d}}$ for segments in $\mathbb{E}^3$ are not transitive. In this section it is shown how to compute a linear extension of $(S, \prec)$ or decide that $(S, \prec)$ contains a cycle. Thus we want to compute an order $o_1, \ldots, o_n$ on the objects in S such that $o_i \prec o_j$ implies $i < j$.

The algorithm that we will give for this problem needs a data structure $\mathcal{D}_\prec$ for storing a subset $S' \subseteq S$, that can return a predecessor in S' of a query object $o \in S$. More formally, $\textsc{Query}(o, \mathcal{D}_\prec)$ returns an object $o' \in S'$ such that $o' \prec o$, or NIL if there is no such object. We call such a query a *predecessor query*. Similarly, we need a structure $\mathcal{D}_\succ$ for *successor queries*. To make our algorithm efficient, the structures should allow for efficient deletions of objects from S' and the preprocessing time should not be too high.

Let us define $\prec_*$ to be the transitive closure of $\prec$. The basis strategy of the algorithm is divide-and-conquer: we pick a pivot object $o_{piv} \in S$, partition the remaining objects into a subset $S_\prec$ of objects o that must come before o_{piv} in the order, because $o \prec_* o_{piv}$, and a subset $S_\succ$ of objects that must come after o_{piv} in the desired order, because $o_{piv} \prec_* o$, and recursively sort these sets. Note that not every pair of objects is comparable under $\prec_*$. Hence, except for the subsets $S_\prec$ and $S_\succ$ there is a third subset $S_\approx$ of objects that cannot be compared to o_{piv} under $\prec_*$. This subset should be sorted recursively as well.

To find the subsets $S_\prec$ and $S_\succ$ efficiently, the data structures $\mathcal{D}_\prec$ and $\mathcal{D}_\succ$ are used. Consider the subset $S_\prec$. By querying $\mathcal{D}_\prec$ with object o_{piv}, we can find an object o such that $o \prec o_{piv}$. We delete o from $\mathcal{D}_\prec$, to avoid that it is reported more than once, and query once more with o_{piv}. Continuing in this manner until the answer to the query is *NIL*, we can find all objects $o \in S$ such that $o \prec o_{piv}$. However, we want to find all objects o such that $o \prec_* o_{piv}$. Thus we also have to query $\mathcal{D}_\prec$ with the objects that we have just found, and query with the new objects that we find, and so forth. Whenever we find an object, it is deleted from $\mathcal{D}_\prec$, and we query with it until we have found all predecessors of it (that have not been found before). This way we can compute the set $S_\prec$ with a number of queries in $\mathcal{D}_\prec$ that is linear in the size of $S_\prec$. Notice that when we find o_{piv} as an answer to a query, then there must be a cycle in the relation. The subset $S_\succ$ can be found in a similar way, using the data structure $\mathcal{D}_\succ$. The subset $S_\approx$ contains the remaining objects.

There is one major problem with this approach: we cannot ensure that the partitioning is balanced, that is, that the sets $S_\prec$, $S_\succ$ and $S_\approx$ have about the same size. Normally, an unbalanced divide-and-conquer algorithm has a quadratic worst-case running time. Fortunately, we can circumvent this if we make the following two observations. First, we note that we need not treat the subset $S_\approx$ separately. We can put the objects of $S_\approx$ into either $S_\prec$ or $S_\succ$, as long as we do it consistently, that is, as long as we put all objects into the same set. It seems that this only makes things worse, because the partitioning gets more unbalanced. But now we observe that it is enough to find the smaller of the two subsets $S_\prec$ and $S_\succ$. The remaining objects—which can be objects of $S_\approx$—are all put into one set. It is possible to find the smaller of the two sets $S_\prec$ and $S_\succ$ with a number of queries that is linear in its size, by doing a 'tandem search': alternatingly, find an object of $S_\prec$ and an object of $S_\succ$, until the computation of one of the two subsets has been completed. Thus we partition S into two subsets in time that is dependent on the the size of the smaller of the two subsets. This means that the more unbalanced the partitioning is, the faster it is performed, leading to a good worst-case running time for the algorithm. There is one problem left that we have not addressed so far: we cannot afford to build the data structures that are needed for the recursive call for the large set from scratch. Fortunately, we can obtain these structures from the structures for S that we already have available, by deleting the elements of the smaller set and the pivot element. (For one of the structures, the elements of the smaller set have already been deleted during the tandem search; for the other structure, these elements still have to be deleted, and we also have to reinsert the elements of the large set that were deleted during the tandem search.)

The algorithm for computing an order on $(S, \prec)$ first builds the data structures $\mathcal{D}_\prec$ and $\mathcal{D}_\succ$ on the set S, and then calls Algorithm 11.1, with the set S and these two data structures as parameters. Algorithm 11.1 outputs a linear extension of $(S, \prec)$ if one exists, and detects a cycle otherwise. The algorithm maintains two queues $\mathcal{Q}_\prec$ and $\mathcal{Q}_\succ$ that store the objects of $S_\prec$ and $S_\succ$ for which we have not yet found all predecessors and successors, respectively. The procedure ENQUEUE adds an object to a queue. Similarly, DEQUEUE deletes an object from the queue. An object o is deleted from the data structure $\mathcal{D}_\prec$ by calling DELETE$(o, \mathcal{D}_\prec)$; a deletion from $\mathcal{D}_\succ$ is performed with a similar call. To delete all objects in a set A, we simply write DELETE$(A, \mathcal{D}_\prec)$.

Algorithm 11.1
Input: A set S of n objects, and the data structures $\mathcal{D}_\prec$ and $\mathcal{D}_\succ$.
Output: A linear extension of $(S, \prec)$ if it exists, and *NIL* otherwise.
1. **if** $|S| = 1$ **then return** S.
2. Make $S_\prec \leftarrow \varnothing$ and $S_\succ \leftarrow \varnothing$, and initialize two empty queues $\mathcal{Q}_\prec$ and $\mathcal{Q}_\succ$.
3. Pick any pivot object $o_{piv} \in S$; ENQUEUE$(o_{piv}, \mathcal{Q}_\prec)$; ENQUEUE$(o_{piv}, \mathcal{Q}_\succ)$.
4. **while** both $\mathcal{Q}_\prec$ and $\mathcal{Q}_\succ$ are not empty
5.   **do** { Compute a new element $o' \in S_\prec$. }
6. $o \leftarrow$ DEQUEUE$(\mathcal{Q}_\prec)$; $o' \leftarrow$ QUERY$(o, \mathcal{D}_\prec)$.
7.   **if** $o' \neq NIL$
8.   **then if** $o' = o_{piv}$
9.   **then return** *NIL*.
10.   **else** ENQUEUE$(o, \mathcal{Q}_\prec)$; ENQUEUE$(o', \mathcal{Q}_\prec)$; DELETE$(o', \mathcal{D}_\prec)$.
11. $S_\prec \leftarrow S_\prec \cup \{o'\}$.
12.   Compute a new element $o'' \in S_\succ$ is a similar way, using $\mathcal{Q}_\succ$ and $\mathcal{D}_\succ$.
13. **if** $\mathcal{Q}_\prec$ is empty (hence, $S_\prec$ is the smaller set)
14.   **then** { Compute the data structures for the recursive calls. }
15.   Restore $\mathcal{D}_\succ$ to the situation before line 4.
16.   DELETE$(S_\prec \cup \{o_{piv}\}, \mathcal{D}_\succ)$; DELETE$(o_{piv}, \mathcal{D}_\prec)$.
17.   Build new predecessor and successor structures $\mathcal{D}'_\prec$ and $\mathcal{D}'_\succ$ for $S_\prec$.
18.   { Sort $S_\prec$ and $S - S_\prec - \{o_{piv}\}$ recursively. }
19.   Call Algorithm 11.1 recursively, with $S_\prec$, $\mathcal{D}'_\prec$ and $\mathcal{D}'_\succ$.
20.   **if** the recursive calls returns *NIL*
21.   **then return** *NIL*
22.   **else** Call Algorithm 11.1 recursively, with $S - \{o_{piv}\} - S_\prec$, $\mathcal{D}_\prec$ and $\mathcal{D}_\succ$.
23.   **if** the recursive calls returns *NIL*
24.   **then return** *NIL*
25.   **else return** the concatenation of $S_\prec$, o_{piv} and $S - \{o_{piv}\} - S_\prec$.
26.   **else** Compute the data structures for the recursive calls as above, reversing the roles of $S_\prec, \mathcal{D}_\prec$ and $S_\succ, \mathcal{D}_\succ$, and sort $S_\succ$ and $S - S_\succ - \{o_{piv}\}$ recursively.

The following lemma proves the correctness of our algorithm.

Lemma 11.2 *Algorithm 11.1 outputs a linear extension of $(S, \prec)$ if one exists, and detects a cycle otherwise.*

Proof: It is straightforward to see that the algorithm never claims to have found a cycle that does not exist. It remains to show that if ORDER outputs a list $o_1, \ldots, o_n$ then this list is a correct order. Assume for a contradiction that $o_i \succ o_j$ for some $i < j$. Then, at some stage of the algorithm, o_i must have been put into $S_\prec$ whereas o_j was put into $S_\succ$, or o_i was put into $S_\prec$ and o_j was the pivot object o_{piv}, or o_i was the pivot object o_{piv} and o_j was put into $S_\succ$. The second and third case both imply that there is a cycle containing o_{piv}, and we can easily verify that the algorithm never fails to discover a cycle containing the pivot object. We thus consider the first case: If $Q_\prec$ is empty after line 12 then all predecessors of o_i have been found, including o_j. Hence, o_j would have been put into $S_\prec$ instead of into $S_\succ$. Similarly, if $Q_\succ$ is empty then o_i would have been put into $S_\succ$. □

Next we prove a bound on the running time of the algorithm. For the sake of simplicity, let us assume that the query time of $\mathcal{D}_\prec$ and the query time of $\mathcal{D}_\succ$ are equal, and let this time be denoted by $Q_\prec(n)$. Similarly, let the time to build these structures on n objects be $T_\prec(n)$, and let $D_\prec(n)$ denote the time for a deletion.

Lemma 11.3 *Algorithm 11.1 runs in $O([T_\prec(n)+n(Q_\prec(n)+D_\prec(n))] \log n)$ time. The running time reduces to $O(T_\prec(n)+n(Q_\prec(n)+D_\prec(n)))$ if $T_\prec(n)/n+Q_\prec(n)+D_\prec(n) = \Omega(n^\alpha)$ for some constant $\alpha > 0$.*

Proof: Since all other operations in the procedure can be done in constant time, the time that we spend is dominated by the operations on the structures $\mathcal{D}_\prec$ and $\mathcal{D}_\succ$. Furthermore, if the size of the smaller of the two subsets $S_\prec$ and $S_\succ$ is m, then we perform at most $2m+2$ queries and deletions on these structures in lines 10 and 12 of the procedure. Restoring a data structure to a situation from the past, which we do in line 15 (or 26), can be done without extra asymptotic overhead if we record all the changes. Finally, we perform m deletions in line 16 (or 26), and we build new data structures for the smaller set. This adds up to $T_\prec(m) + O(1+m)(Q_\prec(n)+D_\prec(n))$ in total for the partitioning.

Next we argue that $m \leqslant n/2$ if the partitioning is successful, that is, if no cycle is found at this point. Suppose that $m > n/2$. Then there must be an object $o \in S_\prec \cap S_\succ$. But this means that o_{piv} will be found as a predecessor or a successor (whichever happens first) and a cycle is detected. Trivially, an unsuccessful partitioning happens at most once, giving a one-time cost of $O(n(Q_\prec(n) + D_\prec(n)))$.

It follows that the total running time $T(n)$ can be bounded by the recursion

$$T(n) \leqslant \max_{0 \leqslant m \leqslant n/2} T_\prec(m) + O(1+m)(Q_\prec(n) + D_\prec(n)) + T(m) + T(n - m - 1),$$

which solves to the claimed time. This can most easily be seen by the following argument: At every partitioning, charge $T_\prec(m)/m + O(1)(Q_\prec(n) + D_\prec(n))$ to each of the m objects in the smaller set and to o_{piv}. We assume that $T_\prec(n)$ is at least linear, so we can bound the charge on a single object by $c(n) := O(T_\prec(n)/n + Q_\prec(n) + D_\prec(n))$.

But every time an object gets charged, the size of the set that contains the object has at least been halved, so the total charge on a single object can be bounded by $c(n) + c(n/2) + c(n/4) + \cdots$. This sums to $O(c(n))$ if $c(n) = \Omega(n^\alpha)$ for some constant $\alpha > 0$, giving a total of $O(nc(n))$. If that is not the case, we can observe that an object gets charged at most $\log n$ times, so we can bound the total time by $O(nc(n) \log n)$.
□

Combining the two lemmas above, we obtain the following theorem.

Theorem 11.4 *Algorithm 11.1 runs in* $O([T_\prec(n) + n(Q_\prec(n) + D_\prec(n))] \log n)$ *time, and outputs an ordered list if* $(S, \prec)$ *does not contain a cycle or finds a cycle otherwise. The running time reduces to* $O\left(T_\prec(n) + n(Q_\prec(n) + D_\prec(n))\right)$ *if* $T_\prec(n)/n + Q_\prec(n) + D_\prec(n) = \Omega(n^\alpha)$ *for some constant* $\alpha > 0$.

Remark 11.5 With a little extra effort, the algorithm can output a witness cycle when $(S, \prec)$ cannot be ordered. To this end we keep track of the successor (predecessor) of each object that we put into $S_\prec$ ($S_\succ$). This extra information enables us to 'walk back' when we find o_{piv} in line 8 or 12 of the algorithm, and report the objects of the cycle.

11.2 Verifying Linear Extensions

In this section it is shown how to verify a given order for a relation $(S, \prec)$. Notice that different orders can be valid for $(S, \prec)$, so it does not suffice to compute a valid order and compare it to the given order. The algorithm uses a straightforward divide-and-conquer approach. It relies on the existence of a data structure $\mathcal{D}_\prec$ for predecessor queries. Unlike in the previous section, however, this data structure need not be dynamic. The algorithm we describe next has as input a list $\mathcal{L}$ of which we have to test whether it corresponds to a valid order.

Algorithm 11.6
Input: A list $\mathcal{L} = \{o_1, \ldots, o_n\}$.
Output: *YES* if $\mathcal{L}$ corresponds to a valid order, *NO* otherwise.
1. **if** $n > 1$
2.   **then** Let $\mathcal{L}_1 = \{o_1, \ldots, o_{\lfloor n/2 \rfloor}\}$ and $\mathcal{L}_2 = \{o_{\lfloor n/2 \rfloor + 1}, \ldots, o_n\}$.
3.   Build a data structure $\mathcal{D}_\prec$ for predecessor queries on $\mathcal{L}_2$.
4.   **for** $i = 1$ **to** $\lfloor n/2 \rfloor$
5.   **do if** QUERY$(o_i, \mathcal{D}_\prec) \neq NIL$
6.   **then return** *NO*.
7.   Call Algorithm 11.6 recursively, with $\mathcal{L}_1$.
8.   Call Algorithm 11.6 recursively, with $\mathcal{L}_2$.
9. **if** at least one of the recursive calls returns *NO*
10.   **then return** *NO*.
11.   **else return** *YES*.

The correctness of the procedure is obvious: If $\mathcal{L}$ does not correspond to a valid order, then, by definition, there are elements o_i, o_j such that $o_i \prec o_j$ and $i > j$. Now either $o_i \in \mathcal{L}_2$ and $o_j \in \mathcal{L}_1$, or $o_i, o_j \in \mathcal{L}_1$, or $o_i, o_j \in \mathcal{L}_2$. The first case is tested by querying with the elements of $\mathcal{L}_1$ in the data structure for predecessor queries on $\mathcal{L}_2$, and the second and third possibility are tested recursively. The following theorem is now straightforward. As before, $T_{\prec}(n)$ denotes the time needed to build the structure $\mathcal{D}_{\prec}$ on a set of n objects, and $Q_{\prec}(n)$ denotes the query time.

Theorem 11.7 *Algorithm 11.6 verifies in $O((T_{\prec}(n) + nQ_{\prec}(n))\log n)$ time whether a list $\mathcal{L}$ corresponds to an order for $(S, \prec)$. The running time of the procedure reduces to $O(T_{\prec}(n) + nQ_{\prec}(n))$ if $T_{\prec}(n)/n + Q_{\prec}(n) = \Omega(n^\alpha)$ for some constant $\alpha > 0$.*

Remark 11.8 Observe that if the procedure reports that $\mathcal{L}$ is not ordered, then it can report a witness pair o_i, o_j of objects such that $i < j$ and $o_j \prec o_i$. If the structure $\mathcal{D}_{\prec}$ is dynamic, then the algorithm can even report all conflicting pairs, as follows. When we test an object $o_i \in \mathcal{L}_1$, we just remove each object $o_j \in \mathcal{L}_2$ that conflicts with o_i from $\mathcal{D}_{\prec}$ and report the pair o_i, o_j, until no more conflicting objects are found. Then we reinsert the objects of $\mathcal{L}_2$ into $\mathcal{D}_{\prec}$, and test the next object of $\mathcal{L}_1$ in the same way. Finally, we find conflicting pairs in $\mathcal{L}_1$ and $\mathcal{L}_2$ recursively.

11.3 Computing and Verifying Depth Orders in Three Dimensions

In this section we apply the general algorithms developed in the previous section to obtain efficient algorithms to compute or verify depth orders in three-dimensional space.

11.3.1 Depth Orders for Line Segments

Let S be a set of n line segments in $\mathbb{E}^3$. We show how to compute or verify depth orders for S in the negative z-direction. The adaptation of the algorithms to depth orders in another direction, or to depth orders with respect to a point, is straightforward. So we want to find or verify a linear extension of the relation $(S, \prec)$, where $s \prec s'$ if segment s is above segment s'. To apply Theorem 11.4, we need dynamic data structures that store a subset $S' \subseteq S$ of segments in $\mathbb{E}^3$ and enable us to find a segment in S' which is above or below a query segment. To this end we hang a curtain from every segment in S'. Now observe that a segment s is below a segment $s' \in S'$ if and only if s intersects the curtain hanging from s'. Thus, we can find a predecessor of s by searching with s for a curtain that intersects s, and report the segment holding that curtain. Such intersection queries with a segment in a set of curtains can be implemented as ray shooting queries: shoot a ray from one of the endpoints of the segment along the segment, and report the first curtain that is hit. If this curtain is intersected by the segment, then we have found a predecessor; otherwise, none of the curtains will be intersected by the segment and the segment does not have a predecessor. Finding a successor of a query segment can be done in a similar way. From Theorem 7.21 we know that ray shooting queries in a set of curtains can be answered in $O(n^{1/3+\varepsilon})$ time, with a structure that has $O(n^{1/3+\varepsilon})$ update time (take $m = n^{4/3}$). Note that the update time is amortized. This is dangerous for us if we restore the data structure to a situation from the past in step 15 of procedure ORDER, since we could perform an expensive deletion too often. We can circumvent this problem by reinserting the elements, instead of restoring the data structure. If we do this, then the amortized update time guarantees us that the total amount of time taken by all the updates is good, leading to the desired running time. If the segments are c-oriented then we can answer queries more efficiently, if we note that the query rays always contain one of the segments. Hence, we can build a separate structure for each of the c possible directions of the query ray. For a fixed direction, we can use Theorem 6.20. The total preprocessing becomes $O(cn \log n)$ and the update time becomes $O(c \log^2 n)$, since we have c separate structures. The query time remains $O(c \log^2 n)$, because we have to query only one of the structures for a given query ray. Combining this with Theorem 11.4 gives us the following result.

Theorem 11.9 *For any $\varepsilon > 0$, it is possible to compute a depth order for a set of n segments in 3-space, or decide that there is cyclic overlap among the segments, in $O(n^{4/3+\varepsilon})$ time. If the segments are c-oriented then the time bound improves to $O(cn \log^3 n)$.*

To verify a given depth order for a set of segments in 3-space, we use the results of Section 11.2. Again, we implement the data structures for predecessor and successor queries as ray shooting structures. This time, however, we do not need dynamic data structures. Hence, we can use Theorem 6.18 for the c-oriented case. We immediately obtain the following theorem.

Theorem 11.10 *For any $\varepsilon > 0$, it is possible to verify a given depth order for a set of n segments in 3-space in $O(n^{4/3+\varepsilon})$ time. If the segments are c-oriented then the time bound improves to $O(cn \log^2 n)$.*

11.3.2 Depth Orders for Triangles

To extend our results to triangles instead of segments, we only need to adapt the data structures for predecessor and successor queries. Let us discuss the structure for predecessor queries; to obtain a structure for successor queries we only have to reverse the roles of 'above' and 'below'.

A triangle t is above another triangle t' if and only if (i) an edge of t is above an edge of t', (ii) t is above a vertex of t', or (iii) a vertex of t is above t'. We already know how to find the triangles t' that satisfy condition (i) for a query triangle t. Next, we show how to handle conditions (ii) and (iii).

First, we consider the case of arbitrary triangles.

Lemma 11.11 *Let P be a set of n points in $\mathbb{E}^3$. There exists a structure that stores P such that a point in P lying above a query triangle t can be reported in $O(n^{1/3+\varepsilon})$ time. The structure uses $O(n^{4/3+\varepsilon})$ storage and has $O(n^{1/3+\varepsilon})$ amortized update time.*

Proof: Project all points in P orthogonally onto the xy-plane. Let $\bar{t}$ be the projection onto the xy-plane of a query triangle t. To find a point above t we select all points whose projection is contained in $\bar{t}$ in a small number of groups. For such a group we can think of t as being a plane, and the question becomes that of reporting a point in a half-space in $\mathbb{E}^3$. The selection can be done using a three-level two-dimensional partition tree: each level filters out those vertices lying on the appropriate side of the line through one of the three edges of $\bar{t}$. Reporting a point in a half-space in $\mathbb{E}^3$ can be done using the half-space emptiness structure of Agarwal et al. [5]. This structure uses $O(n^{4/3+\varepsilon})$ preprocessing, and has $O(n^{1/3+\varepsilon})$ amortized update time. Because the preprocessing and the query time are essentially determined by the least efficient level, we obtain a total structure with $O(n^{1/3+\varepsilon})$ query time and $O(n^{1/3+\varepsilon})$ update time. See Section 2.6 for further details on the analysis of (multi-level) partition trees. $\square$

Clearly we can use this structure to handle condition (ii). The structure for condition (iii) is the same (up to some dualizations) as the structure for (ii) that we just described. We conclude that a dynamic structure for predecessor queries in a set of triangles exists with $O(n^{1/3+\varepsilon})$ query and $O(n^{1/3+\varepsilon})$ update time, using $O(n^{4/3+\varepsilon})$ preprocessing time and space. As before, these bounds are amortized, implying that we should reinsert the elements in step 15 of procedure ORDER, instead of restoring the data structure to a situation from the past.

For the c-oriented case we refer the reader to Section 16.5, where we show that a vertex below a c-oriented query triangle can be found in $O(\log^3 n)$ time with a structure that uses $O(c^2 n \log^2 n)$ storage and has $O(c^2 \log^3 n)$ update time. Furthermore, we note that finding a vertex below an axis-parallel rectangle can be done slightly more efficient: by using a range tree with a priority search tree as associated structure we obtain a structure whose query and update time are $O(\log^2 n)$. See Section 16.4. This enables us to order axis-parallel rectangles rectangles more efficiently than c-oriented triangles. To handle condition (iii), we have to find a triangle below a query vertex v. Note that a triangle below a query vertex can be found by shooting a ray from the vertex into a fixed direction. By Theorem 6.16, we can answer these queries in $O(c^2 \log^2 n \log \log n)$ time with a structure that uses $O(n \log^2 n)$ storage and has $O(c^2 \log^2 n \log \log n)$ update time.

The above combined with Theorem 11.4 leads to the following result.

Theorem 11.12 *For any $\varepsilon > 0$, it is possible to compute a depth order for a set of n triangles in 3-space, or decide that there is cyclic overlap among the triangles, in $O(n^{4/3+\varepsilon})$ time. If the triangles are c-oriented then the time bound improves to $O(c^2 n \log^4 n)$, and if the objects are axis-parallel rectangles then the algorithm takes $O(n \log^3 n \log \log n)$ time.*

We now turn our attention to the verification of a given depth order for a set of triangles. In the algorithm of Section 11.2 we have to test whether the triangles in a list $\mathcal{L}_1$ do not lie above any triangle in a list $\mathcal{L}_2$. Testing whether there is an edge of a triangle in $\mathcal{L}_1$ that passes above an edge of a triangle in $\mathcal{L}_2$ can be done in $O(n^{3/2+\varepsilon})$ time in the general case, and in $O(cn \log n)$ time in the c-oriented case, as we have seen when we studied segments. To test for conflicts corresponding to condition (ii), we build a structure on the triangles in $\mathcal{L}_1$ that reports the first triangle that is hit by a query ray from infinity into the negative z-direction. Next, we shoot rays towards each vertex of all triangles in $\mathcal{L}_2$; when we know the first triangle that is hit by the ray towards a certain vertex, we can decide if there is any triangle above the vertex. We know that ray shooting from infinity into a fixed direction is equivalent to ray shooting from a fixed point. Hence, we can use the results of Chapter 5, where we have shown that we can answer such ray shooting queries in $O(n^{1/3+\varepsilon})$ time after $O(n^{4/3+\varepsilon})$ preprocessing in the general case (Theorem 5.13), and in $O(c^2 \log n)$ time after $O(n \log n)$ preprocessing in the c-oriented case (Theorem 5.7). Hence, in $O(n^{4/3+\varepsilon})$ and $O(c^2 n \log n)$ time, respectively, we can decide if there is a vertex of a triangle in $\mathcal{L}_2$ that is below some triangle in $\mathcal{L}_1$. To test condition (iii) we build a similar structure on the triangles in $\mathcal{L}_2$ (only this time for query rays into the positive z-direction), and we query with vertices of triangles in $\mathcal{L}_1$. This leads to the following theorem.

Theorem 11.13 *For any $\varepsilon > 0$, it is possible to verify a given depth order for a set of n triangles in 3-space in $O(n^{4/3+\varepsilon})$ time. If the triangles are c-oriented then the time bound improves to $O(c^2 n \log^2 n)$.*

Remark 11.14 In the c-oriented case, we can trade an $O(c)$-factor in the running time for an $O(\log n)$ factor, by using Theorem 6.15 instead of Theorem 5.7. This yields a total running time of $O(cn \log^3 n)$.

11.3.3 Depth Orders for Polygons

Consider the case where we want to compute a depth order for a set of polygons in 3-space, instead of a set of triangles. Let n be the total number of vertices of the polygons. First, we triangulate every polygon, which can be done in $O(n)$ time in total [17]. Observe that one polygon is above another polygon if and only if one of the triangles in the triangulation of the first polygon is above one of the triangles of the second polygon. Hence, we can use the same data structures as before to find predecessors and successors. However, if the polygons do not have constant complexity then there is a slight problem: the triangles that correspond to the same polygon must stay together in the order, so when we find one triangle as a predecessor or successor we have to report the other triangles as well. This is problematic, because the number of other triangles can be large. Suppose that during our tandem search we suddenly have to add a very large polygon to one of the subsets; if we find out in the next step that the other subset is complete, then we have spent a lot of time that we cannot charge to the smaller subset. An elegant solution to this problem can be obtained if we realize that we can choose any particular pivot object we like. Hence, we can choose the polygon with the largest complexity as pivot object. The tandem search for the sets $S_\prec$ and $S_\succ$ now proceeds as follows. We find successors and predecessors using the data structures for triangles. However, when we find a large polygon for, say, $S_\prec$, we first allow $S_\succ$ to catch up. Thus we search for successors until the complexity of $S_\succ$—that is, the total number of vertices of all polygons in $S_\succ$—is greater than the complexity of $S_\prec$. When this happens, we start querying for predecessors again, and so on and so forth, until one of the subsets is completed. This way the extra work that we have to do, caused by adding a large polygon to what turns out to be the larger set, is bounded by the time spent on one polygon. Since the pivot polygon is chosen to be the largest polygon in the set, we can charge this extra work to the pivot polygon. Clearly, each polygon is charged at most once this way. Thus the asymptotic running time of the algorithm remains the same, and we have the following theorem.

Theorem 11.15 *For any $\varepsilon > 0$, it is possible to compute a depth order for a set of polygons in 3-space with n vertices in total, or decide that there is cyclic overlap among the polygons, in $O(n^{4/3+\varepsilon})$ time. If the polygons are c-oriented then the time bound improves to $O(c^2 n \log^4 n)$, and if the polygons are axis-parallel then the algorithm takes $O(n \log^3 n \log \log n)$ time.*

The adaptation of the verification procedure to polygons is fairly straightforward, and we leave it as an (easy) exercise to the reader.

Theorem 11.16 *For any $\varepsilon > 0$, it is possible to verify a given depth order for a set of polygons in 3-space with n vertices in total in $O(n^{4/3+\varepsilon})$ time. If the polygons are c-oriented then the time bound improves to $O(c^2 n \log^2 n)$.*

Chapter 12

Conclusions

In this part of the book we studied the computation of depth orders, which are needed to remove hidden surfaces using the painter's algorithm. In the planar case, we presented a data structure that can answer depth order queries for a set S of m polygons with n vertices in total: Given a query direction $\vec{d}$, the data structure computes in $O(m)$ time a depth order on S for direction $\vec{d}$. The data structure uses $O(m)$ storage. Perhaps the most interesting result that we achieved in this part concerns depth orders in three-dimensional space. We gave the first algorithms that compute or verify a depth order for a set of n line segments (or polygons) in three-dimensional space in subquadratic time. The running time of our algorithms depends on the type of objects we want to sort. In Table 12.1 we summarize our results for computing depth orders for sets of line segments in three-dimensional space. In this table, n denotes the number of line segments and ε is a constant that can be chosen arbitrarily small. The results can be extended to polygons. This sometimes entails an extra logarithmic factor in the running time; see Theorems 11.15 and 11.16.

	axis-parallel	c-oriented	arbitrary
computation	$n \log^3 n$	$cn \log^3 n$	$n^{4/3+\varepsilon}$
verification	$n \log^2 n$	$cn \log^2 n$	$n^{4/3+\varepsilon}$

Table 12.1: Results on computing depth orders for segments in three-dimensional space.

The obvious open problem is to improve these bounds. For the axis-parallel and c-oriented case it might be possible to shave off some logarithmic factors. In the general case, there is more room for improvement, but we believe that it will be difficult to give an algorithm with $o(n^{4/3})$ running time. See also the remarks that we make in the concluding chapter of Part D on the complexity of computing visibility maps.

Another important topic for further research is to find efficient ways to deal with cyclic overlap. The algorithms that we have presented can detect cyclic overlap, but it is still open how to remove the overlap by cutting the objects into (not too many) smaller pieces. It would be very useful if one could find a minimal number of pieces that is necessary to remove all cyclic overlap. However, computing a minimal number of cuts seems to be a hard problem, so one might want to concentrate on finding a number of cuts that is only a constant factor from the optimum. See Chazelle et al. [21], who study related problems.

Part D

Computing Visibility Maps

Chapter 13

Introduction

The *visibility map* of a set of objects in three-dimensional space is the subdivision of the viewing plane into maximal connected regions such that in each region one object in seen, or no object is seen. (See Section 13.2 for a more detailed discussion of the structure of visibility maps.) The output of an object space hidden surface removal algorithm should be a combinatorial representation of this visibility map—for example, in the form of a doubly connected edge list [104]—where each region is labeled with the face that is visible inside that region, or with *NIL* if none of the faces is visible in that region. In this chapter we present algorithms to compute such a combinatorial representation of the visibility map of a set of polyhedra in $\mathbb{E}^3$.

13.1 Output-Sensitive Hidden Surface Removal

Many object-space hidden surface removal algorithms have been developed for computing the visibility map of a set S of polyhedra in $\mathbb{E}^3$ with n edges in total. Almost all of these algorithms only deal with non-intersecting polyhedra. In that case, the regions in the visibility map are bounded by parts of the projections of edges of the polyhedra, and the vertices of the map are either intersections between two projected edges or they are projections of vertices of the polyhedra. Early object-space methods compute the visibility map as follows: they project the edges onto the viewing plane, compute all their intersections, and determine which intersection points are visible. Crude implementations of this approach run in time $O(n^2)$ [47, 83]. More careful implementations [62, 90, 110] run in time $O((n+i)\log n)$ or $O(n\log n+i+j)$, where i denotes the number of intersections between the projected edges and j is the number of intersections between the projected polygons. The problem with these methods is that they are insensitive to the output size of the problem. That is, if the visibility map has k edges, we would prefer an algorithm whose running time depends on k, so that when k is small the algorithm becomes more efficient. We call such an algorithm *output-sensitive*. Note that it is possible that k is very small—even a constant—while i is quadratic in n. Hence, none of the above-mentioned techniques is output-sensitive. Mulmuley [87, 88] gives a 'quasi-output-sensitive' hidden surface removal method; its running time is a sum of weights associated with all intersections of the projected edges, where the weight of an intersection decreases as the number of objects hiding it increases. Still, also this method might require quadratic time to produce a trivial output. (It should be mentioned here that his method is the only one to handle intersecting polyhedra and objects with curved boundaries.)

In 1984 Güting and Ottmann [68] (see also [69]) were the first to present an output-sensitive algorithm, for the special case of c-oriented polygons that are parallel to the viewing plane. Their algorithm runs in $O(c^2(n + k) \log^2 n)$ time. It roughly works as follows. The polygons are treated one by one, in order of increasing distance to the viewpoint. Hence, when a new polygon is processed, all the others that can hide it already have been treated. The algorithm maintains the shadow of the processed polygons, that is, the union of their projections onto the viewing plane. Note that the part of the new polygon that is visible is exactly the part whose projection lies outside the current shadow. See Figure 13.1. Thus the problem can be solved if

Figure 13.1: Processing a new polygon.

the shadow can be stored in a dynamic data structure that allows us to compute its intersection with the projection of the new polygon. Güting and Ottmann presented a data structure for c-oriented shadows with $O(c^2 \log^2 n + k')$ query time, where k' is the number of intersections of the new rectangle with the current shadow, and $O(c^2 \log^2 n)$ update time, leading to an algorithm with the stated running time. This result has been improved and generalized in a number of ways. A special case that has received considerable attention [14, 62, 63, 106] is hidden surface removal in a set of horizontal axis-parallel rectangles—also called the *window rendering problem*. The best result obtained so far is due to Bern [14] and Goodrich, Overmars and Atallah [63]; their algorithms run in time $O((n + k) \log n)$. Recently, Preparata, Vitter and Yvinec [107] have studied the slightly more general case of axis-parallel blocks in space for which a depth order on the faces exists and is known. They show how to compute the visibility map with respect to an arbitrary viewpoint in $O((n + k) \log n \log \log n)$ time. Another special case for which an efficient output-sensitive algorithm has been given is that of a polyhedral terrain: Reif and Sen [108] present an algorithm with $O((n + k) \log n \log \log n)$ running time. The most general output-sensitive solutions have been proposed by Overmars and Sharir [97, 98, 115]. They show how to compute the visibility map of a set of n horizontal triangles viewed from a point at $z = \infty$ in time $O(n\sqrt{k} \log n)$. They also describe a second, more complicated, algorithm with a running time of $O(n^{4/3} \log^{2.89} n + n^{4/5+\varepsilon} k^{3/5})$. Using new results on dynamic partition trees, this has recently been improved by Agarwal and Sharir [6] to $O(n^{1+\varepsilon} + n^{2/3+\varepsilon}k^{2/3})$.[1] Finally, Katz et al. [72] propose an efficient algorithm for a set of fat objects for which a depth order is known. They achieve a running time of $O((C(n) + k)\log^2 n)$, where $C(n) \geqslant n$ is the maximum complexity of the union of the projection of certain subsets of the objects. For polyhedral terrains—where $C(n) = O(n\alpha(n))$—they are able to shave off a log-factor, thus obtaining an $O((n\alpha(n) + k)\log n)$ algorithm and improving the result of Reif and Sen.

Many of the above mentioned algorithms follow the approach of Güting and Ottmann. Thus the objects are processed in order of increasing distance to the view point—in other words, according to a depth order—and their shadow is maintained. Also the methods that follow a different approach rely strongly on the fact that a depth order on the objects exists and is known. The restriction of the availability of a depth order is a severe one. Even in a simple set of axis-parallel blocks cyclic overlap can occur at many places. See Figure 13.2 for an example. Finding a method for com-

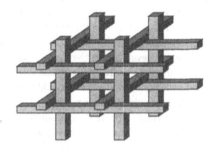

Figure 13.2: Blocks with cyclic overlap.

puting visibility maps in an output-sensitive manner that can deal with cyclic overlap was considered an important open problem in computational geometry [19, 125]. In the next chapter we present such an algorithm. The method uses some of the ray shooting data structures presented in Part B. For non-intersecting axis-parallel and c-oriented polyhedra the algorithm runs in $O((n+k)\log n)$ and $O(c^3 n \log n + kc^2 \log n)$ time, respectively, and non-intersecting arbitrary polyhedra take $O(n^{1+\varepsilon} + n^{2/3+\varepsilon}k^{2/3})$ time. The polyhedra are allowed to have holes. Both parallel and perspective views can be computed. Notice that these algorithms are at least as fast as the previously known algorithms which only work when a depth order exists. The method can even be extended to deal with intersecting polyhedra, as is shown in Chapter 15. For axis-parallel and c-oriented polyhedra, this entails only a logarithmic increase in the time bound; for arbitrary polyhedra, the increase is somewhat larger. These results are taken from [41, 42, 43].

In practice, for example in animation, the scene that one wants to display often changes over time. In Chapter 16 we therefore study the problem of maintaining

---

[1]In fact, the running time of their algorithm is sometimes somewhat better, depending on a parameter $\zeta \leqslant k$.

the visibility map of a set of polyhedra under insertions into or deletions from the set. Of course, one can recompute the whole visibility map from scratch, whenever a polyhedron is added or deleted. But this takes a lot of time, even when there are only few (or no) changes in the map. Hence, it seems natural to try and maintain the visibility map in a more clever way. Strangely enough, this problem has not received much attention. There are only two papers on dynamic hidden surface removal that we know of. Bern [14] considers the case where the scene consists of n axis-parallel rectangles that are parallel to the viewing plane. He obtains a solution that is output-sensitive: the time needed to insert or delete a rectangle is $O(\log^2 n \log \log n + k \log^2 n)$, where k is the number of changes in the visibility map. Cheng [30] considers the more general case of horizontal polygons, but his solution is intersection-sensitive instead of output-sensitive: updates take time $O(\sqrt{n} \log^{1.5} n + i \log n + k)$ where i is the total number of intersections between the inserted or deleted polygon and the other polygons in the scene. Note that i can be $\Omega(n)$, even for a polygon that is not visible at all. Moreover, Cheng considers the hidden line problem instead of the hidden surface problem, that is, he does not maintain the visibility map as a collection of regions, but he only maintains the visible portions of the boundary of each polygon. Again, both methods rely on the fact that a depth order exists and is known.

In Chapter 16 we use the ideas developed in the static case to maintain the visibility map of a set of non-intersecting polyhedra when cyclic overlap is allowed. For the axis-parallel and c-oriented case, this gives a quite efficient solution: the time needed to insert or delete a polyhedron of constant complexity is $O(\log^2 n \log \log n + k \log^2 n)$ and $O(c^4(k + 1) \log^3 n)$, respectively. For arbitrary polyhedra the method is—as to be expected—not as fast: the update time is $O((k + 1)(n + k_S)^{1/3+\epsilon})$, where k_S is the combinatorial complexity of the total visibility map. Observe, however, that this is still more efficient than recomputing the whole map from scratch for a large range of values of k and k_S.

13.2   The Structure of the Visibility Map

Before we describe our algorithms, let us give a more precise definition of the visibility map. For a viewing direction $\vec{d}$, we define the visibility map of S in direction $\vec{d}$, denoted by $\mathcal{M}_{\vec{d}}(S)$, as follows. A point q is called *visible* (in direction $\vec{d}$ with respect to S) if and only if the ray from infinity into direction $\vec{d}$ towards q does not intersect any polyhedron in S before it intersects q. Let h be a plane that is orthogonal to $\vec{d}$. To obtain $\mathcal{M}_{\vec{d}}(S)$, we project all visible points on the faces of the polyhedra in S orthogonally onto h; this subdivides h into maximal connected regions where either one unique face is visible or no face at all is visible. This subdivision, where each region is labeled with the face that is visible inside that region or with *NIL* if no face is visible, is the visibility map. To avoid confusion with the edges and vertices of the polyhedra in the scene, we call the edges of the visibility map *arcs*, and the vertices of the map *nodes*.

We can also define visibility maps for perspective views. In that case, we define a point q to be visible from the view point p if and only if the interior of the line segment $\overline{pq}$ connecting p and q does not intersect any polyhedron in S. To define the visibility map of S from p we can make two choices: we can take a viewing plane or a viewing sphere. Note that we can 'unfold' the visibility map on, for example, the southern hemisphere to obtain the visibility map on the plane that is tangent to the south pole of the sphere. Similarly, we can combine the visibility maps on the planes that are tangent to the north pole and the south pole into the visibility map on the sphere. Hence, the choice of definition does not really matter. Since we will describe the algorithms for parallel views, we need not go into this any further.

To simplify the notation, we describe the algorithms in the remainder of this chapter for parallel views into the negative z-direction, and we take the viewing plane to be the xy-plane. Intuitively, we look at the scene from above. The adaptation to parallel views into other directions is trivial. The adaptation to perspective views is straightforward as well. (There is one thing that requires a little attention: if the view point is inside the convex hull of of the scene, then it may happen that none of the vertices of the polyhedra is visible. This is a problem for our algorithms, which can be avoided by splitting the scene with a plane through the viewpoint.) Because the viewing direction is fixed and the set S is usually clear from the context, we often write $\mathcal{M}$ instead of $\mathcal{M}_{\vec{d}}(S)$.

Next, we take a closer look at the the structure of the visibility map of a set S of polyhedra. A first observation we can make is that we do not need to consider faces whose face normal points downward. Such faces are 'on the back side' of the polyhedra—hence, they are called *back faces*—and thus cannot be visible. Trivially, the back faces can be determined in linear time in total. Let F be the remaining set of faces of the polyhedra in S, and let E and V be the set of edges and vertices of the faces in F. The following observation characterizes the arcs and nodes of $\mathcal{M}$. Here the projections are orthogonal projections onto the xy-plane; for an object o, we denote this projection by $\bar{o}$.

Observation 13.1 *An arc a in $\mathcal{M}$ is of one of the following two types:*

(A1) $a \subseteq \bar{e}$ *(a part of a projected edge)*
(A2) $a \subseteq \overline{f_1} \cap \overline{f_2}$ *(a part of the the projection of the intersection of two faces)*

A node ν in $\mathcal{M}$ is of one of the following five types:

(N1) $\nu = \bar{v}$ *(a projected vertex)*
(N2) $\nu = \overline{e_1} \cap \overline{e_2}$ *(the intersection of two projected edges)*
(N3) $\nu = \bar{e} \cap \bar{f}$ *(the projection of the intersection of an edge and a face)*
(N4) $\nu = \bar{e} \cap \overline{f_1} \cap \overline{f_2}$ *(the intersection of a projected edge and the projection of*
the intersection of two faces)
(N5) $\nu = \overline{f_1} \cap \overline{f_2} \cap \overline{f_3}$ *(the projection of the common intersection of three faces)*

where $v \in V$, $e, e_1, e_2 \in E$ and $f, f_1, f_2, f_3 \in F$.

See Figure 13.3 for an illustration. We say that an arc (or node) of $\mathcal{M}$ *is defined by* the faces and/or edges (edges or vertex) involved in its definition. It will be convenient to be able to refer to the point on one of the polyhedra that is visible and whose projection is a certain point q on the viewing plane. We denote this point by *lift*(q).

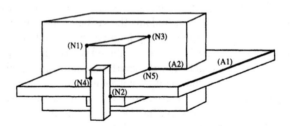

Figure 13.3: The different types of nodes and arcs in a visibility map.

For a node ν of type (N1), for example, $lift(\nu)$ is its defining vertex. We extend this definition to arcs of the visibility map in the natural way: $lift(a) = \{lift(q) : q \in a\}$. Thus $lift(a)$ is the part of its defining edge (or the part of the intersection of its two defining faces) whose projection is a. One final piece of terminology. The face that is visible inside a region $\mathcal{R}$ of $\mathcal{M}$ is called the *background face* of $\mathcal{R}$, denoted $bf(\mathcal{R})$. If no face is visible inside $\mathcal{R}$, then we define $bf(\mathcal{R}) = NIL$.

We close this section with a few remarks about the maximum combinatorial complexity—that is, the total number of nodes, arcs, and regions—of the visibility map $\mathcal{M}$ of a set of polyhedra with n vertices in total. Observe that if the polyhedra are non-intersecting, then we only have arcs of type (A1), and nodes of types (N1) and (N2). From this it easily follows that the maximum combinatorial complexity of the visibility map is $O(n^2)$. By considering a set of long thin axis-parallel blocks whose projection forms a grid-like pattern, we see that this bound is tight. For intersecting polyhedra the situation is not as simple, because of the nodes of types (N4) and (N5). For example, there can be $\Theta(n^3)$ intersections between triples of faces, but not all of them can show up as nodes in the visibility map. Indeed, the maximum complexity of the visibility map of a set of intersecting polyhedra with n vertices in total is not much larger than in the non-intersecting case: Pach and Sharir [99] prove an $O(n^2\alpha(n))$ bound on this complexity and show that it is tight.

Chapter 14

Non-Intersecting Polyhedra

This chapter considers hidden surface removal in a set of non-intersecting polyhedra. In the first section we give a general output-sensitive algorithm that computes the visibility map $\mathcal{M}$ of a set S of non-intersecting polyhedra with n vertices in total. In three later sections, we apply this general method to obtain efficient algorithms for axis-parallel, c-oriented and arbitrary polyhedra. In fact, the algorithm as stated only outputs the arcs of the map, but it is easily extended to report, for example, a doubly connected edge list of $\mathcal{M}$ where each region is labeled with its background face.

14.1 The Algorithm

Recall that we assumed for notational convenience that the viewing direction is the negative z-direction and that the viewing plane is the xy-plane. The algorithm moves a vertical sweep line l (that is, a sweep line parallel to the y-axis) from left to right over the viewing plane. The sweep line halts at the nodes of $\mathcal{M}$ and also at the projection of every vertex in V. To simplify the description, we assume that no two event points have equal x-coordinate. This assumption can easily be removed by using the lexicographical order, instead of an order on x-coordinate, in the sequel. The algorithm maintains the invariant that at any time during the sweep the part of $\mathcal{M}$ to the left of l has been computed. More precisely, all the nodes to the left of or on the sweep line have been computed, and also all arcs that are incident to these nodes. The computation of $\mathcal{M}$ is finished when l sweeps over the rightmost node of the map.

Let us study how to advance the sweep line. Suppose the sweep line reaches a node ν of the map. The arcs that are incident to ν and lie to the left of it—we call these arcs the *incoming arcs* of ν—have already been computed. Hence, in order to maintain the invariant we only have to compute the incident arcs that lie to the right of ν—called the *outgoing arcs* of ν. When the sweep line reaches the projection of a vertex, we also have to decide whether this projected vertex is a node of $\mathcal{M}$, that is, whether the vertex is visible. Only when this is the case we have to compute its outgoing arcs. This is illustrated in Figure 14.1. In this figure, only those edges of $\mathcal{M}$ are shown that have been computed at this stage in the sweep. In Figure 14.1(a) we have to compute two new edges of the visibility map. After this has been done,

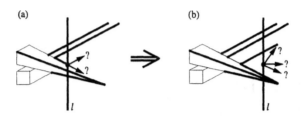

Figure 14.1: Advancing the sweep line.

we are allowed to advance the sweep line to the next event point, which happens to be a projected vertex. See Figure 14.1(b). Because this vertex is visible, we have to compute the (in this case three) outgoing arcs of the corresponding node.

Deciding whether a projected vertex $\bar{v}$ is a node of the map is done as follows. The algorithm maintains a binary search tree $\mathcal{T}$ on the arcs that intersect the sweep line at its current position. See Figure 14.1, where these edge are drawn fat. The order in which the arcs are stored in the tree, is the order in which they are encountered when we walk from top to bottom along the sweep line. With each arc, we store the face which is visible in the region below (that is, in the negative y-direction of) the arc. Hence, by searching with the projected vertex in $\mathcal{T}$, we can find in $O(\log n)$ time the face that is visible in the region that contains $\bar{v}$. If v lies in front of this face, then $\bar{v}$ is a node of $\mathcal{M}$. Otherwise v is hidden and it does not define a node of the map.

It remains to show how to compute the outgoing arcs of a node ν of $\mathcal{M}$. We know the edges (or the vertex) that define(s) ν. Hence, we also know the edges defining its outgoing arcs, and the question becomes to compute the right endpoint of such an arc a. Let $e(a) \in E$ be the edge that defines a. Clearly, the right endpoint μ of a is either the right endpoint of $\overline{e(a)}$—as in Figure 14.2(a)—or it is the intersection of $\overline{e(a)}$ and $\bar{e}$ for some other edge $e \in E$. In the latter case, we can make a further distinction: either

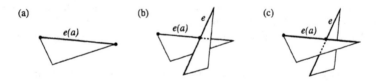

Figure 14.2: The three possibilities for the right endpoint of an arc.

$e(a)$ is below e—as in Figure 14.2(b)—or e is below $e(a)$—as in Figure 14.2(c). Note that if $e(a)$ is below e, then $e(a)$ becomes invisible at the intersection point, because it is hidden by the face incident to e. Similarly, e becomes invisible (or visible) when it is below $e(a)$.

First, let us see how we can detect the moment at which $e(a)$ becomes invisible.

For an edge $e \in E$, we define $curt(e)$ to be the curtain hanging from e. Thus $curt(e)$ is the set of points that are hidden by e, and the curtains hanging from the edges of a face bound the region in space that is hidden by that face. Hence, $e(a)$ becomes invisible as soon as it intersects a curtain. So we can detect the moment at which $e(a)$ becomes invisible by ray shooting in the set $curt(E) = \{curt(e) : e \in E\}$ with a ray $\rho(a)$ which is defined as follows: $\rho(a)$ is the ray along $e(a)$, starting at the point on $e(a)$ that projects onto ν, and directed to the right. See Figure 14.3.

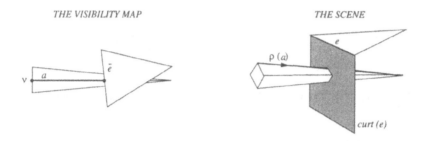

Figure 14.3: The right endpoint of a is the projection of $\rho(a) \cap curt(e)$.

The second case, where an edge e disappears below the face that is incident to $e(a)$, is more problematic. One might think that we can find edges disappearing below $e(a)$ by hanging curtains *upside down* from the edges. However, some of the edges below $e(a)$ were never visible in the first place, and the fact that they are below the face incident to $e(a)$ does not change anything. Somehow we have to avoid reporting such edges. Fortunately, there is an elegant solution to this problem, which makes use of a ray $\hat{\rho}(a)$ defined as follows. Note that if $e(a)$ is incident to two faces in F (as in the case in Figure 14.3, for example) then none of the edges below $e(a)$ can be visible, because F only contains front faces. For reasons to become apparent later, we define $\hat{\rho}(a) = \rho(a)$ when $e(a)$ is incident to two faces. Now consider the more interesting case where $e(a)$ is incident to exactly one face in F. Let $f \in F$ be the face that is directly below $e(a)$ at the point whose projection is ν. In other words, of the two faces that are visible in the two regions bounded by a, f is the one that is not incident to $e(a)$. Note that edges which are below f will not be visible. We define $\hat{\rho}(a)$ to be the projection of $\rho(a)$ onto f. (If there is no face below $e(a)$, then we define $\hat{\rho}(a)$ to be the projection of $\rho(a)$ onto a plane which is below all the faces in the scene.) See Figure 14.4. By definition of $\hat{\rho}(a)$, for any edge that is below $e(a)$ and is also visible, the curtain hanging from this edge must be intersected by $\hat{\rho}(a)$. Thus we can determine the first edge disappearing below $e(a)$ by ray shooting with $\hat{\rho}(a)$ in the set $curt(E)$. Note that any curtain that intersects $\rho(a)$ also intersects $\hat{\rho}(a)$. Hence, shooting with $\hat{\rho}(a)$ in $curt(E)$ also takes care of the first case, where $e(a)$ itself disappears. This leads to the following lemma.

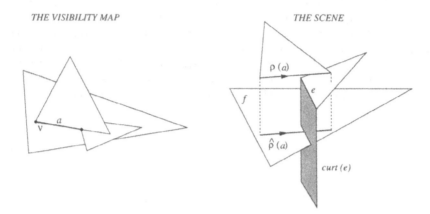

Figure 14.4: The right endpoint of a is the projection of $\rho(a) \cap curt(e)$.

Lemma 14.1 *The right endpoint μ of a is the leftmost point of the following event points:*

- *the projection of the right endpoint of $e(a)$*

- *the projection of the first intersection of $\hat{\rho}(a)$ with a curtain in $curt(E)$*

Proof: Suppose μ is not the projection of the right endpoint of $e(a)$. By Observation 13.1, μ must be the intersection of $\overline{e(a)}$ with the projection $\bar{e}$ of some $e \in E$. From the definition of $\hat{\rho}(a)$ it follows that e must be above $\hat{\rho}(a)$: if it is below $\hat{\rho}(a)$ then the intersection point cannot be visible. (In the case where $\hat{\rho}(a) = \rho(a)$ this intersection point is hidden by the two faces incident to $e(a)$; if $\hat{\rho}(a) \neq \rho(a)$ it is hidden by the face f containing $\hat{\rho}(a)$.)

It remains to show that the projection of the first intersection point q of $\hat{\rho}(a)$ with a curtain in $curt(E)$ is visible. Suppose for a contradiction that q is hidden by some face f'. From the definition of q it follows that f' must lie above $\hat{\rho}(a)$. Because $curt(e)$ is the *first* curtain that is intersected, we know that $\hat{\rho}(a)$ cannot intersect a curtain hanging from an edge of f' before it intersects $curt(e)$. But since f' is above q, and the polyhedra are non-intersecting, this means that f' must also be above the starting point of $\hat{\rho}(a)$, contradicting either the definition of $\hat{\rho}(a)$ or the fact that ν is visible. $\square$

So we know how to compute the right endpoint μ of an arc a of $\mathcal{M}$: perform a ray shooting query with $\hat{\rho}(a)$ in the set $curt(E)$, and compare the projection of the right endpoint of $e(a)$ to the projection of the first intersection point of $\hat{\rho}(a)$ with a curtain; whichever of the two is closer to ν is the endpoint μ that we seek.

But before we can state the final algorithm, we have to deal with two subtleties that we have swept under the rug so far. First, the computation of $\hat{\rho}(a)$. Since we know the edge defining a, we have the ray $\rho(a)$ readily available. However, to compute

$\hat{\rho}(a)$ we need the face that is immediately below the starting point of $\rho(a)$. This face is not always known. An easy way to find it is to shoot a ray into the negative z-direction towards the starting point of $\rho(a)$. The first face whose interior is hit by this ray is the face that we need. In fact, we also need this face to update the tree $\mathcal{T}$, because it will be a face that is visible in a region bounded by a. A second fine point is the following. A node has, in general, more than one incoming arc, and we want to avoid storing the node more than once as an event point. We could store all the event points in a search structure, which allows us to test whether a node has been computed before. However, there is a simpler and more elegant solution to this problem: store a node that has been computed as the right endpoint of arc a if and only if the edge $e(a)$ defining a is also visible to the right of the node. Because there is exactly one such edge for each node which is not a visible vertex and no such edge for visible vertices—which are already inserted into the queue at the start of the algorithm—this solves our problem. We thus arrive at the following algorithm.

Algorithm 14.2
Input: A set S of non-intersecting polyhedra.
Output: The edges of $\mathcal{M}(S)$.
1. Remove backfaces from the polyhedra in S. Let F be the set of remaining faces, and let E and V be the set of edges and vertices of the faces in F.
2. Build a data structure $\mathcal{D}(F)$ for ray shooting from infinity into the negative z-direction in the set F.
3. Build a data structure $\mathcal{D}(curt(E))$ for ray shooting in the set $curt(E)$.
4. Sort the set of projections of the vertices in V by x-coordinate, and store them in a priority queue $\mathcal{Q}$.
5. Initialize a binary search tree $\mathcal{T}$.
6. **while** $\mathcal{Q}$ is not empty
7.   **do** Remove the point ν with smallest x-coordinate from $\mathcal{Q}$.
8.   **if** ν is the projection of a vertex in V
9.   **then** test whether ν is a node of $\mathcal{M}$ by searching in $\mathcal{T}$
10.   **if** ν is a node of $\mathcal{M}$
11.   **then for** each outgoing arc a_i of ν
12.   **do** Compute $\hat{\rho}(a_i)$ using $\mathcal{D}(F)$.
13.   Query with $\hat{\rho}(a_i)$ in $\mathcal{D}(curt(E))$ and compute the right endpoint μ of a according to Lemma 14.1.
14.   **if** $e(a)$ is visible to the right of μ
15.   **then** Insert μ into $\mathcal{Q}$.
16.   Insert $a_i = \overline{\nu\mu}$ into $\mathcal{T}$ and report a_i as an edge of $\mathcal{M}$.
17.   Delete the incoming arcs of ν from $\mathcal{T}$.

Although Lemma 14.1 shows that our algorithm computes the right endpoint of an arc correctly, it does not prove that every arc of the map is eventually computed. This is proved in the next lemma.

Lemma 14.3 *Algorithm 14.2 outputs every edge of the visibility map exactly once.*

Proof: If we can prove that the left endpoint of every arc a of $\mathcal{M}$ is inserted into $\mathcal{Q}$ exactly once, then the lemma follows by Lemma 14.1. The argument is by induction on the x-coordinates of the nodes. If the left endpoint ν of a is the projection of a visible vertex v, then ν is inserted into Q in line 4 of the algorithm. Clearly, none of the edges incident to v is visible both to the left and to the right of v, so ν will not be inserted in line 15. If ν is not the projection of a visible vertex, then it must be the intersection of the projection of two edges e and e'. One of those edges, say e', disappears behind the face that is incident to the other edge; hence, e' cannot be visible on both sides of ν. The other edge e does not disappear, and is therefore visible on both sides of ν. Thus e defines an incoming arc a' of ν, and ν is inserted when a' is computed; the fact that a' has been computed follows by induction.   □

Before we state the final result, let us summarize what we need to implement the algorithm described above. Suppose we can

(i) build in $B_F(n)$ time a data structure $\mathcal{D}(F)$ for ray shooting queries from infinity into the negative z-direction in the set F of polygonal faces, which has $Q_F(n)$ query time and uses $M_F(n)$ storage,

(ii) build in $B_{curt(E)}(n)$ time a data structure $\mathcal{D}(curt(E))$ for ray shooting queries in the set $curt(E)$ of curtains, which has $Q_{curt(E)}(n)$ query time, and uses $M_{curt(E)}(n)$ storage.

If these conditions are fulfilled then we have the following theorem.

Theorem 14.4 *The visibility map of a set of non-intersecting polyhedra with n vertices in total can be computed in time*

$$O(\ n\log n + B_F(n) + B_{curt(E)}(n) + k(\log n + Q_F(n) + Q_{curt(E)}(n))\ ),$$

using $O(M_F(n) + M_{curt(E)}(n))$ storage, where k is the combinatorial complexity of the visibility map.

Proof: The correctness of Algorithm 14.2 has been proved in Lemma 14.3, so it remains to analyze its time and storage requirements. As for the time complexity, we observe that lines 1–5 take $O(n\log n + B_F(n) + B_{curt(E)}(n))$ time. In the remainder of the algorithm, we spend $O(\log n)$ time per vertex to test if it is visible, we spend $O(Q_F(n) + Q_{curt(E)}(n))$ time per arc of the map to compute its right endpoint, and $O(\log n)$ per arc to update Q and T. The time bound follows. To prove the bound on the amount of storage, we note that the number of arcs that the sweep line intersects at any given moment is linear. Hence, the size of T and of Q is $O(n)$. The storage bound follows, since we may assume that the amount of storage used by the ray shooting structures is at least linear.   □

14.2 Axis-Parallel Polyhedra

To apply the result of the previous section to axis-parallel polyhedra we need two data structures: one for ray shooting from infinity into a fixed direction in a set of axis-parallel polyhedra, and one for ray shooting in a set of curtains hanging from axis-parallel edges. We already noted in Part B that ray shooting from infinity into a fixed direction is equivalent to ray shooting from a fixed point. Hence, we can apply Theorem 5.3, which states that ray shooting queries from a fixed point in a set of axis-parallel polyhedra with n vertices in total can be answered in $Q_F(n) = O(\log n)$ time, with a structure that uses $B_F(n) = M_F(n) = O(n \log n)$ storage and preprocessing time. To answer ray shooting queries in a set of axis-parallel curtains, we can use data structures from Chapter 7. However, we can do better, because we only shoot along edges of the polyhedra and along the projection of these edges onto faces of the polyhedra. Hence, the number of distinct directions of the query ray is only six. Thus we can build six different structures, one for each possible direction of the query ray. If the direction of the query ray is fixed, then ray shooting queries in axis-parallel curtains can be answered in $O(\log n)$ time after $O(n \log n)$ preprocessing, according to Theorem 6.17. We obtain the following theorem.

Theorem 14.5 *The visibility map of a set of non-intersecting axis-parallel polyhedra with n vertices in total can be computed in time $O((n + k) \log n)$ using $O(n \log n)$ storage, where k is the combinatorial complexity of the visibility map.*

14.3 *c*-Oriented Polyhedra

To implement the data structures for the hidden surface removal algorithm in the case of *c*-oriented polyhedra, we use the same two observations as in the axis-parallel case. The first observation is that ray shooting from infinity into a fixed direction is equivalent to ray shooting from a fixed point. Hence, we can use a structure from Chapter 5 to implement $\mathcal{D}(F)$. In particular, we can apply Theorem 5.7, which states that ray shooting queries from a fixed point in a set of *c*-oriented polyhedra with n vertices in total can be answered in $Q_F(n) = O(c^2 \log n)$ time with a structure that uses $B_F(n) = M_F(n) = O(n \log n)$ storage and preprocessing time. Observe that if $c = \Omega(\log n)$ it is better to use Theorem 6.15 instead of Theorem 5.7, to achieve $O(c \log^2 n)$ query time instead of $O(c^2 \log n)$. The second observation concerned the structure $\mathcal{D}(curt(E))$ for ray shooting in curtains: we only shoot along edges of the polyhedra and along the projection of these edges onto faces of the polyhedra, which implies that the number of distinct directions of the query ray is limited. For *c*-oriented polyhedra, the edges have c different orientations, and the projections of edges onto faces have $O(c^3)$ different orientations. We build $O(c^3)$ different structures, one for each possible direction of the query ray. For a fixed direction of the query ray we apply Theorem 6.18, which states that ray shooting queries in *c*-oriented curtains can be answered in $O(c \log n)$ time after $O(n \log n)$ preprocessing. It follows that the ray shooting queries in $curt(E)$ can be answered in time $Q_{curt(E)} = O(c \log n)$, with a structure that uses $B_{curt(E)} = M_{curt(E)} = O(c^3 n \log n)$ storage and preprocessing time. This leads to the following theorem.

Theorem 14.6 *The visibility map of a set of non-intersecting c-oriented polyhedra with n vertices in total can be computed in time*

$$O(c^3 n \log n + kc^2 \log n) \qquad \text{if } c = O(\log n)$$
$$O(c^3 n \log n + kc \log^2 n) \qquad \text{otherwise}$$

using $O(c^3 n \log n)$ storage, where k is the combinatorial complexity of the visibility map.

14.4   Arbitrary Polyhedra

To implement Algorithm 14.2 for arbitrary (non-intersecting) polyhedra, we can apply Theorem 5.13 and Theorem 7.21, which state that ray shooting queries from a fixed point and ray shooting queries in a set of curtains, respectively, can be answered in $O(n^{1+\varepsilon}/\sqrt{m})$ time after $O(m^{1+\varepsilon})$ preprocessing, for any fixed $\varepsilon > 0$ and any $n \leqslant m \leqslant n^2$. Hence, the running time of the algorithm is $O(m^{1+\varepsilon} + k(n^{1+\varepsilon}/\sqrt{m}))$. It only remains to decide upon the value of m. To minimize the running time, we would like to have $m^{1+\varepsilon} = k(n^{1+\varepsilon}/\sqrt{m})$. To achieve this (approximately), we must choose $m = n^{2/3} k^{2/3}$. This holds provided that $n^{2/3} k^{2/3}$ is at least n; otherwise we set $m = n$. This way we can obtain for any $\varepsilon > 0$ a running time of $O(n^{1+\varepsilon} + n^{2/3+\varepsilon} k^{2/3})$. Unfortunately, we do not now the value of k in advance, so how can we choose m to be a function of k? We can solve this problem in the same way as in [97]. We 'guess' a constant value $k' = k_0$ for k, choose m accordingly, and run the algorithm. As soon as we detect that $k > k'$ we stop the algorithm, multiply k' by two, and try again. The algorithm halts when $k \leqslant k'$ for the first time. Due to our multiplication scheme, we know that $k' < 2k$, so the total running time is bounded by

$$O(n^{1+\varepsilon} + n^{2/3+\varepsilon} k_0^{2/3}) + O(n^{1+\varepsilon} + n^{2/3+\varepsilon}(2k_0)^{2/3}) + \cdots + O(n^{1+\varepsilon} + n^{2/3+\varepsilon}(2k)^{2/3}) \quad =$$
$$O(n^{1+\varepsilon}(\log(k/k_0) + 1)) + O(n^{2/3+\varepsilon} k_0^{2/3}) \sum_{i=0}^{\log(k/k_0)+1} 2^{2i/3} \quad =$$
$$O(n^{1+\varepsilon} \log n) + O(n^{2/3+\varepsilon} k^{2/3}).$$

This leads to the following theorem.

Theorem 14.7 *For any $\varepsilon > 0$, it is possible to compute the visibility map of a set of non-intersecting polyhedra with n vertices in total in $O(n^{1+\varepsilon} + n^{2/3+\varepsilon} k^{2/3})$ time using $O(n^{1+\varepsilon} + n^{2/3+\varepsilon} k^{2/3})$ storage, where k is the combinatorial complexity of the visibility map.*

Note that both for small (near-constant) values of k, as well as for large (near-quadratic) values of k, the algorithm is close to optimal.

Chapter 15

Intersecting Polyhedra

In this chapter we extend the results of the previous chapter to intersecting polyhedra. Again, we first describe the general algorithm and then we implement it in the specific cases of axis-parallel, c-oriented and arbitrary polyhedra.

15.1 The Algorithm

When we try to extend the algorithm of the previous chapter to intersecting polyhedra, we encounter two difficulties. The first problem is that Lemma 14.1 only accounts for nodes of types (N1) and (N2), and not for the three other types of nodes that can occur in a scene of intersecting polyhedra. The second problem is the following. In the non-intersecting case the left endpoint of an arc is either a projected vertex or the right endpoint of another arc. Using this we can prove that every left endpoint is inserted into the event queue $\mathcal{Q}$ and, hence, that every arc is computed during the sweep. In the intersecting case this is no longer true: nodes of type (N3) can have outgoing edges without having incoming edges. Therefore we cannot use the sweep line approach. In the remainder of this section we show how to deal with these two problems.

First, we extend Lemma 14.1 to intersecting polyhedra. Assume that one endpoint ν of an arc a is known, and that we want to compute the other endpoint μ. We define rays $\rho(a, \nu)$ and $\hat{\rho}(a, \nu)$ is a similar manner as we defined $\rho(a)$ and $\hat{\rho}(a)$ in the non-intersecting case. If arc a is defined by an edge $e(a)$—a is of type (A1)—then $\rho(a, \nu)$ is a ray along $e(a)$ starting at the point on $e(a)$ that projects onto ν and directed towards μ. (More precisely, its projection is directed towards μ.) Furthermore, we define the ray $\hat{\rho}(a, \nu)$ in the same way as before: if $e(a)$ is incident to two faces then $\hat{\rho}(a, \nu) = \rho(a, \nu)$, otherwise $\hat{\rho}(a, \nu)$ is the projection of $\rho(a, \nu)$ onto the face immediately below $\text{lift}(\nu)$. If arc a is defined by the intersection $f_1 \cap f_2$ of two faces—a is of type (A2)—then $\rho(a, \nu)$ is the ray along this intersection, also starting at the point that projects onto ν and directed towards μ. In this case, we always have $\hat{\rho}(a, \nu) = \rho(a, \nu)$. The analog of Lemma 14.1 is now as follows. The reader is encouraged to find examples of each of the given event points in Figure 13.3 on page 154.

Lemma 15.1 *The other endpoint μ of a is the point closest to ν of the following event points:*

- *the projection of an endpoint of $e(a)$ (if a is of type (A1))*

- *the projection of the first intersection of $\hat{\rho}(a, \nu)$ with a curtain in $curt(E)$*

- *the projection of an endpoint of $f_1 \cap f_2$ (if a is of type (A2))*

- *the projection of the first intersection of $\rho(a, \nu)$ with a face in F*

- *the projection of the first intersection of $\hat{\rho}(a, \nu)$ with a face in F*

Proof: First we show by a simple case study that all different types of vertices that are mentioned in Observation 13.1 are included in the five event points above. Then we show that the event point closest to ν must be visible, thus completing the proof.

If $\mu = \overline{v}$ for a vertex $v \in V$, then v is obviously an endpoint of $e(a)$, which is the first event point. If $\mu = \overline{e} \cap \overline{e'}$ for two edges $e, e' \in E$, then μ is the projection of the first intersection of $\hat{\rho}(a, \nu)$ with a curtain in $curt(E)$, which corresponds to the second event point; this follows in exactly the same way as in the non-intersecting case. If $\mu = \overline{e \cap f}$ for an edge $e \in E$ and a face $f \in F$, then there are two cases: we are shooting along e, which is handled by the fourth event point, or we are shooting along the intersection of two faces (namely f and the face containing e), which is handled by the third event point. If $\mu = \overline{e} \cap \overline{f_1 \cap f_2}$ for an edge $e \in E$ and faces $f_1, f_2 \in F$, then again we have two cases: we are shooting along $f_1 \cap f_2$, which corresponds to the second event point, or we are shooting along e which is handled by the last event point. Finally, if $\mu = \overline{f_1 \cap f_2 \cap f_3}$, then μ is an event point of the fourth type: we are shooting along the intersection of two of the faces and μ corresponds to the point where the ray intersects the third face.

It remains to show that the first point μ of the five event points must be visible. Suppose for a contradiction that μ is hidden by some face $f' \in F$. Let q be the point on $\hat{\rho}(a, \nu)$ whose projection is μ. If μ is hidden by f', then f' must be above q. Moreover, it follows from the definition of μ that the part of $\hat{\rho}(a, \nu)$ between its starting point and point q does not intersect f' or any curtain hanging from an edge of f'. This contradicts either the fact that ν is a node of $\mathcal{M}$, or the definition of $\hat{\rho}(a, \nu)$. $\square$

Next, we turn our attention to the second problem, which is the fact that we cannot use a sweep line approach. Instead, we use the following strategy, which was first described by Overmars and Sharir [97]. Let $\mathcal{C}$ be a connected component of the planar graph that corresponds to $\mathcal{M}$. The idea is that if we can find (at least) one node of $\mathcal{C}$ we can discover the rest of $\mathcal{C}$ arc by arc, using Lemma 15.1. To find one node in every connected component, we compute which of the vertices of the polyhedra are visible. The projection of such a vertex is a node of $\mathcal{M}$. The next lemma shows that this gives us at least one node on every connected component. In the non-intersecting case this is readily seen, because the leftmost node of every connected component must be the projection of a visible vertex. If the polyhedra are allowed to intersect this is no longer true, however, and the proof is more involved.

Lemma 15.2 *Every connected component of $\mathcal{M}$ contains at least one node that is the projection of a visible vertex.*

Proof: Consider any component $\mathcal{C}$ of $\mathcal{M}$. Observe that $\mathcal{C}$ is enclosed in a unique region $\mathcal{R}$ of $\mathcal{M}$. Let $bf(\mathcal{C}) = bf(\mathcal{R})$ be the background face of this region; if $bf(\mathcal{R}) = NIL$, that is, no face is visible $\mathcal{R}$, then the leftmost vertex of $\mathcal{C}$ must be a visible vertex and we are done. We associate with $\mathcal{C}$ a set $S(\mathcal{C})$ of points in the viewing plane: $S(\mathcal{C})$ contains all the nodes of $\mathcal{C}$, all the points on arcs of $\mathcal{C}$, and all the points inside regions whose outer boundary is formed by arcs of $\mathcal{C}$. Recall that, for a point q in the viewing plane, $lift(q)$ is defined to be the point on one of the polyhedra that is visible and whose projection is q. For each point $q \in S(\mathcal{C})$, consider the vertical distance of $lift(q)$ to $bf(\mathcal{C})$, that is, the length of the segment which is parallel to the z-direction (the viewing direction) and connects $lift(q)$ to $bf(\mathcal{C})$. Let $q_{max} \in S(\mathcal{C})$ be the point such that this distance is maximal. (If this point is not unique, we take the one that is maximal in the lexicographical order.) We will prove that q_{max} must be a visible vertex.

Let us assume for simplicity that no other faces in the scene are parallel to $bf(\mathcal{R})$. (If this is not the case, then we have to formulate the argument a bit more carefully, using the lexicographical order.) Obviously, point q_{max} cannot be in the interior of a region of $\mathcal{C}$. Now consider a plane h through q_{max} that is parallel to $bf(\mathcal{C})$. By definition of q_{max}, all lifted points in $S(\mathcal{C})$ are below h. This immediately disqualifies points in the interior of arcs of $\mathcal{C}$, and nodes of types (N2) and (N4) as possibilities for q_{max}: the edges containing these lifted points would intersect h and define an arc of $\mathcal{C}$, and the lifted points on this arc have greater distance to $bf(\mathcal{C})$ than q_{max}. Hence, the remaining possibilities for q_{max} are nodes of types (N1), (N3) and (N5). Suppose for a contradiction that q_{max} is of type (N3), that is, $q_{max} = e \cap f$. By definition of q_{max}, we approach h when we move along the visible part of e towards q_{max}. After intersecting h at point q_{max}, e is above h. But then there is a point on f above h. We claim that the projection of this point belongs to $S(\mathcal{C})$, thus contradicting the definition of q_{max}. To prove this claim, we note that $bf(\mathcal{C}) \neq f$. Hence, q_{max} must be on the outer boundary of the region where f is visible, implying that the points inside this region belong to $S(\mathcal{C})$. We conclude that q_{max} cannot be of type (N3). Using a similar argument, we can show that q_{max} cannot be of type (N5), leaving only the possibility that q_{max} is of type (N1). $\qquad\square$

Lemma 15.2 proves the correctness of our new strategy, which is to compute the visible vertices first, and then to compute the components which are attached to the corresponding nodes. To decide whether a vertex v is visible, we can use the data structure for ray shooting into a fixed direction from infinity: shoot a ray towards v, and check whether v is hidden by the first face in F whose interior is hit. Note that in the second stage of the algorithm, where we trace the connected components using Lemma 15.1, we need a ray shooting structure that we did not have in the non-intersecting case: we have to shoot with the rays $\rho(a, \nu)$ and $\hat{\rho}(a, \nu)$ in F. (In the non-intersecting case, we only needed a structure for shooting from infinity into a fixed direction in F.) Another extra data structure that we need comes from the fact that we have to avoid reporting arcs more than once. In the sweep line approach we always shoot from left to right, making a local test possible to see if a node has to be inserted into the event queue Q. In our new strategy this is no longer possible. Hence, we implement the event queue Q as a binary search tree; with each node of M that is stored in Q we also store which of its incident arcs already have been computed. We thus obtain the following algorithm.

Algorithm 15.3
Input: A set S of possibly intersecting polyhedra.
Output: The edges of $M(S)$.
1. Remove backfaces. Let F, E and V be the sets of remaining faces, edges and vertices, respectively.
2. Build a data structure $\mathcal{D}(F)$ for ray shooting from infinity into the negative z-direction on the set F.
3. Build a data structure $\mathcal{D}(curt(E))$ for ray shooting in the set $curt(E)$.
4. Build a data structure $\mathcal{D}'(F)$ for ray shooting in the set F.
5. Shoot a ray from infinity into the viewing direction towards every vertex $v \in V$, to test if v is visible.
6. Store the visible vertices in an event queue Q, implemented as a binary search tree.
7. **while** Q is not empty
8.   **do** Remove any node ν from Q.
9.   **for each** incident arc a_i of ν that has not been computed yet
10.   **do** Compute $\rho(a_i, \nu)$, and $\hat{\rho}(a_i, \nu)$ using $\mathcal{D}(F)$.
11.   Query with $\rho(a_i, \nu)$ and $\hat{\rho}(a_i, \nu)$ in $\mathcal{D}(F)$ and $\mathcal{D}'(F)$ and compute the other endpoint μ of a_i according to Lemma 15.1.
12.   Insert μ into Q if necessary.
13.   Record that arc a_i has been computed for μ.
14.   Report a_i as an edge of M.

Let us summarize what is needed to implement the above algorithm. Suppose we can

(i) build in $B_F(n)$ time a data structure $\mathcal{D}(F)$ for ray shooting queries from infinity into the negative z-direction in the set F of polygonal faces, which has $Q_F(n)$ query time and uses $M_F(n)$ storage,

(ii) build in $B_{curt(E)}(n)$ time a data structure $\mathcal{D}(curt(E))$ for ray shooting queries in the set $curt(E)$ of curtains, which has $Q_{curt(E)}(n)$ query time, and uses $M_{curt(E)}(n)$ storage.

(iii) build in $B'_F(n)$ time a data structure $\mathcal{D}'(F)$ for ray shooting queries in the set F of polygonal faces, which has $Q'_F(n)$ query time and uses $M'_F(n)$ storage,

(It seems that we do not need $\mathcal{D}(F)$, since we also have $\mathcal{D}'(F)$. However, $\mathcal{D}(F)$ is queried $O(n+k)$ times and $\mathcal{D}'(F)$ only $O(k)$ times. Hence, if $k = o(n)$ it pays to have a separate structure $\mathcal{D}(F)$—provided that this structure has a better performance than $\mathcal{D}'(F)$, of course.)

The time complexity of the algorithm readily follows from the description, leading to the following theorem.

Theorem 15.4 *The visibility map of a set of possibly intersecting polyhedra with n vertices in total can be computed in time*

$$O(\ B_F(n) + B'_F(n) + B_{curt(E)}(n) + nQ_F(n) + k(\log n + Q_F(n) + Q'_F(n) + Q_{curt(E)}(n))\ ),$$

using $O(M_F(n) + M'_F(n) + M_{curt(E)}(n) + k)$ storage, where k is the combinatorial complexity of the visibility map.

15.2 Axis-Parallel Polyhedra

To implement Algorithm 15.3 for axis-parallel polyhedra, we need three data structures. Two of these, $\mathcal{D}(F)$ and $\mathcal{D}(curt(E))$, were also needed in the non-intersecting case. We implement them in the same way, using Theorem 5.3 and Theorem 6.17, respectively. The extra data structure $\mathcal{D}'(F)$ is used for 'arbitrary' ray shooting queries in F. Observe that we only shoot along edges, along projections of edges onto other faces, and along the intersections of faces. Hence, the query rays have only six different directions, like in the non-intersecting case. For each direction we build a separate structure, using either Theorem 6.3 or Theorem 6.6 depending on k. We obtain the following theorem.

Theorem 15.5 *For any $\varepsilon > 0$, it is possible to compute the visibility map of a set of possibly intersecting axis-parallel polyhedra with n vertices in total in time*

$$
\begin{array}{ll}
O(n \log^2 n + k \log n (\log \log n)^2) & \text{if } k = O(n^{1+\varepsilon}) \\
O(n^{1+\varepsilon} + k \log n) & \text{otherwise}
\end{array}
$$

where k is the combinatorial complexity of the visibility map. The algorithms use $O(n \log n + k)$ and $O(n^{1+\varepsilon} + k)$ storage, respectively. The running time of the first algorithm is randomized.

15.3 c-Oriented Polyhedra

The implementation of the data structures that Algorithm 15.3 needs in the case of c-oriented polyhedra is as follows. The data structures $\mathcal{D}(F)$ and $\mathcal{D}(curt(E))$ were also needed in the non-intersecting case. We can implement them in the same way, using Theorem 5.7 (or Theorem 6.15, depending on c) and Theorem 6.18, respectively. The extra data structure $\mathcal{D}'(F)$ is used for 'arbitrary' ray shooting queries in F. We noticed before that we only shoot along edges, along projections of edges onto other faces, and along the intersections of faces. Hence, like in the non-intersecting case, the query rays have a bounded number of distinct directions. This number is dominated by the number of different orientations of the intersections between two faces, which is $O(c^4)$. For each direction we build a separate structure, using either Theorem 6.10 or Theorem 6.11 or Theorem 6.15, depending on k and c. Note that the increase of the number of possible directions of the query rays from $O(c^3)$ to $O(c^4)$ also influences the performance of $\mathcal{D}(curt(E))$. We obtain the following theorem.

Theorem 15.6 *For any $\varepsilon > 0$, it is possible to compute the visibility map of a set of possibly intersecting c-oriented polyhedra with n vertices in total in time*

$$O(c^4 n \log^2 n + kc^2 \log n (\log \log n)^2) \quad \text{if } c = O(\log n/(\log \log n)^2) \text{ and } k = O(n^{1+\varepsilon})$$
$$O(c^4 n^{1+\varepsilon} + kc^2 \log n) \qquad\qquad \text{if } c = O(\log n) \text{ and } k = \Omega(n^{1+\varepsilon})$$
$$O(c^4 n \log^2 n + kc \log^2 n) \qquad\qquad \text{otherwise}$$

where k is the combinatorial complexity of the visibility map. The algorithms use $O(c^4 n \log n + k)$, $O(c^4 n^{1+\varepsilon} + k)$, and $O(c^4 n \log^2 n + k)$ storage, respectively. The running time of the first algorithm is randomized.

15.4   Arbitrary Polyhedra

To implement Algorithm 15.3 for arbitrary polyhedra, we need one extra data structure, as compared to the non-intersecting case. This data structure $\mathcal{D}'(F)$ must answer
ray shooting queries in arbitrary polyhedra with arbitrary rays. We use Theorem 7.27
which states that one can obtain $O(n^{1+\varepsilon}/m^{1/4})$ query time after $O(m^{1+\varepsilon})$ preprocessing. As before, we want to choose m such that the running time is minimal. This is
achieved when $m = kn/m^{1/4}$, so when $m = n^{4/5}k^{4/5}$. By guessing the value of k in
the same way as in the non-intersecting case, we can thus trace the components in
$O(n^{4/5+\varepsilon}k^{4/5})$ randomized time.

Recall that we have to compute which vertices are visible before we can start
tracing the components. This is done by shooting rays from infinity towards every
vertex. Hence, the structure $\mathcal{D}(F)$ is queried n times, before the tracing process
begins. According to Theorem 5.13, we can implement $\mathcal{D}(F)$ such that its query time
is $O(n^{1/3+\varepsilon})$ and its preprocessing time is $O(n^{4/3+\varepsilon})$. Hence, the visible vertices can
be found in $O(n^{4/3+\varepsilon})$ time in total. For the second phase of the algorithm, where we
trace the components, we choose the parameter m in the same way as for the other
structures, namely dependent on k. We obtain the following theorem.

Theorem 15.7 *For any $\varepsilon > 0$, it is possible to compute the visibility map of a set of possibly intersecting polyhedra with n vertices in total in $O(n^{4/3+\varepsilon} + n^{4/5+\varepsilon}k^{4/5})$ time using $O(n^{4/3+\varepsilon} + n^{4/5+\varepsilon}k^{4/5})$ storage, where k is the combinatorial complexity of the visibility map.*

Chapter 16

Dynamization

In this chapter we present an algorithm for maintaining the visibility map of a set of non-intersecting[1] polyhedra, under insertions into and deletions from the set. The algorithm is closely related to the static algorithms of the previous chapters. However, some additional data structures are needed, for example to find visible vertices of new components that become visible after a deletion. In the first section we give an overview of the method, describing more precisely how the visibility map is represented and which additional data structures we need to support updates. Next, we discuss in two separate sections the algorithms for inserting and deleting polyhedra. Then we show how to implement the algorithm in the specific cases of axis-parallel, c-oriented and arbitrary polyhedra.

16.1 Overview of the Method

Before we describe the algorithms for inserting and deleting polyhedra, we must be a bit more specific about the way $\mathcal{M}(S)$ is represented and about what the output of our algorithms should be. Recall that we assumed that the viewing direction is the negative z-direction and that the viewing plane is the xy–plane. Let F, E and V be the sets of faces, edges and vertices of the polyhedra in S, respectively, with back faces removed. To simplify the discussion, we augment F with a large face $f_{-\infty}$ that is below everything in the scene and always remains present. Hence, the visibility map is a collection $R = \{\mathcal{R}_1, \ldots, \mathcal{R}_s\}$ of regions that form a subdivision of the viewing plane such that in each region one face of a polyhedron is visible or $f_{-\infty}$ is visible. We define A and N to be the set of arcs and nodes of the regions of $\mathcal{M}(S)$, respectively. However, these arcs (nodes) are lifted to the background face of the corresponding region. Observe that, since every node ν of the visibility map is incident to at least two regions, there are at least two points corresponding to ν in N. Moreover, some of these points can be identical; hence, N is actually a multi-set. Note also that $lift(\nu)$—the point on a polyhedron in S that projects onto ν and is visible—is one of the points in N that correspond to a node ν. The same comments apply to A.

---

[1]For axis-parallel and c-oriented polyhedra the algorithms can be adapted such that they also handle intersecting polyhedra. See [39].

Each region is bounded by a number of *boundary cycles*, which are cyclic chains of arcs of A; one of these cycles constitutes the outer boundary of the region, the other cycles form the boundary of holes in the region. We store each cycle in a concatenable queue (implemented e.g. as a 2-3 tree [8]). The order in which we store the cycle corresponds to a traversal of the cycle such that the corresponding region lies to the right. This means that the outer boundary is traversed in clockwise order, whereas the boundaries of the holes are traversed in counterclockwise order. We also have crosspointers between each cycle and the face that is visible in the region that it bounds. This is closely related to the representation of the visibility map by Bern [14]. Using the concatenable queue representation, we can split cycles or concatenate two chains to form a cycle in logarithmic time. (Because we have to choose a point on each cycle as a starting point, one of the chains will be delivered in two pieces after the split. These pieces then have to be merged.)

Let $\mathcal{P}$ denote the polyhedron of constant complexity that we want to insert into or delete from S. Thus, given $\mathcal{M}(S)$, we want to compute $\mathcal{M}(S \cup \{\mathcal{P}\})$ or $\mathcal{M}(S - \{\mathcal{P}\})$, respectively. The number of changes in the view, denoted k, is equal to the number of changes in R, A and N. Note that an old region disappears and a new one appears if the background face of the region changes. Thus, if the face that we see through a 'hole' in the scene changes, then we are allowed to spend an amount of time that is dependent on the complexity of the hole. In the remainder of this chapter we prove that these changes in the view can be reported in an output-sensitive manner. The algorithm that achieves this needs structures for ray shooting queries in polyhedra, and in curtains. It also needs data structures for the following range searching query:

- Report the leftmost point—or, all points—of a given set of points in $\mathbb{E}^3$ below (or above) a query face of constant complexity.

More precisely, we need the following six structures.

$\mathcal{D}_1$: A structure for ray shooting queries from infinity into the negative z-direction on the set F.

$\mathcal{D}_2$: A structure for ray shooting on the set $curt(E)$ of curtains hanging downwards from the edges in E.

$\mathcal{D}_3$: A structure for answering range queries as defined above on the set V.

$\mathcal{D}_4$: A structure for ray shooting on the set $upcurt(A)$ of curtains 'hanging' upwards from the lifted arcs in A.

$\mathcal{D}_5$: A structure for answering range queries as defined above on the set N.

Of course, the structures must be dynamic. Let $Q_i(m)$, $1 \leqslant i \leqslant 5$, be the query time[2] of structure $\mathcal{D}_i$ when it stores m objects, let $U_i(m)$ be its update time, and $M_i(m)$ the amount of storage it uses. We will prove the following theorem.

---

[2] The query time of the range searching structures $\mathcal{D}_2$ and $\mathcal{D}_5$ actually depends not only on n but also on the number of answers to the query.

Theorem 16.1 *The visibility map of a set of polyhedra with n vertices in total can be maintained in time*

$$O(\ k \log n + Q_1(n) + (k+1)Q_4(k_S) + Q_5(k_S) + \sum_{1 \leqslant i \leqslant 3} U_i(n) + k \sum_{4 \leqslant i \leqslant 5} U_i(k_S)\ )$$

per insertion of a polyhedron of constant complexity, and time

$$O(\ (k+1)(\sum_{1 \leqslant i \leqslant 3} Q_i(n) + Q_5(k_S)) + \sum_{1 \leqslant i \leqslant 3} U_i(n) + k \sum_{4 \leqslant i \leqslant 5} U_i(k_S)\ )$$

per deletion of a polyhedron of constant complexity, where k is the number of changes in the view and k_S is the combinatorial complexity of the visibility map. The method uses $O(\sum_{1 \leqslant i \leqslant 3} M_i(n) + \sum_{4 \leqslant i \leqslant 5} M_i(k_S))$ storage.

The bound on the amount of storage used by the method is immediate. The time bounds are proved in the next two sections. In the first of these sections we show that insertions can be performed in the stated bound and in the second section this is shown for deletions.

16.2 Insertions

The algorithm to insert a polyhedron $\mathcal{P}$ is as follows.

Algorithm 16.2
Input: A polygon $\mathcal{P}$.
Output: The updated visibility map $\mathcal{M}(S \cup \{\mathcal{P}\})$.
1. **for** every front face[3] f of $\mathcal{P}$
2.   **do** Report all arcs that are (partially) below f.
3.   Remove the portions of these arcs from the affected boundary cycles.
4.   Compute all the new arcs.
5.   Assemble the new cycles from the new arcs and the affected cycles.
6.   Update all data structures.

Below a more detailed description of the steps of the algorithm is given.

Consider the insertion of one face f of $\mathcal{P}$. In step 2 the hidden arcs are found. An arc $a \in A$ is hidden by f if and only if at least one of the following two cases occurs: an endpoint of a is below f, or a is below an edge of f. Thus we can find all the arcs that are (partially) hidden by f as follows. First, we perform a query with f in $\mathcal{D}_5$ to find all the nodes in N below f. This takes $Q_5(k_S)$ time. Second, we perform repeated ray shooting queries along each edge e of f in the set $upcurt(A)$, as follows. Shoot a ray along e from one of its endpoints. If the first curtain that is hit is intersected by e, then report it and shoot another ray along e from the intersection point. Repeat this process until all curtains intersected by e have been found. This takes $O((k+1)Q_4(k_S))$ time in total. Once we have computed in step 2 the arcs that are (partially) hidden, we split them into visible and invisible parts. This is easily done because f has constant complexity. The invisible parts are discarded, and we are left with the visible subchains of each affected cycle. We conclude that steps 2 and 3 take $(k + 1)Q_4(k_S) + Q_5(n)$ time in total.

In step 4 we have to compute the new arcs that appear because of the insertion of face f. These arcs are of two types: (i) the projection of an edge of f onto some other face, or (ii) the projection of an edge of some other face onto f.

Let us first consider the arcs of type (i). Let a be such an arc and let e be the edge of f that defines a. The endpoints of a correspond to either a vertex of e, or the intersection of the projection of e and the projection of some $e' \in E$. To find the first type of endpoints we shoot rays from infinity into the negative z-direction towards both vertices of e. Now vertex v of e is visible—and therefore an endpoint of a new arc—if and only if the first face that is hit lies below v. Thus we can find these endpoints in $Q_1(n)$ time using $\mathcal{D}_1$. Next we turn our attention to the endpoints that are the intersection of the projection of e and the projection of some $e' \in E$. In fact, we do not have to do much to find these endpoints, because we already computed the relevant information in step 2. Namely, if the intersection of the projections of e and e' is visible, then e must be above the face which is below e'. Hence, the projection of e' onto the face below it must have contributed an arc to A that is below e. (It is, of course, possible that not only the projection of e' onto the face below it, but also e' itself is below e. In this case we have found *two* arcs of A in step 2 that ensure that we discover the visible intersection between e and e'.) Conversely, each arc of A below e induces an endpoint of a new arc. Thus finding the endpoints of new arcs which are the intersection of a projected edge of f with another projected edge amounts to simply looking at the arcs of A below e which have been reported in step 2. After having found these endpoints we simply pair them up to obtain the new arcs.

The next type of arcs that we have to find are the arcs of type (ii), which are the projection of other edges onto f. Fortunately, these are easy to determine using the information from step 2. Consider the projection of the visible part of some edge e' onto f. Clearly, before the insertion of f this part must have been projected onto some other face(s) that are now hidden by f. In other words, these parts contributed arcs to A that are now hidden by f and have thus been found in step 2. Hence, to find the arcs of type (ii) we have to check for each arc of A that was reported in step 2 whether it was the projection of an edge that is above f. If this is the case then the projection of this edge onto f gives us a new arc of type (ii). Note that we may have to merge some arcs into one.

Summarizing, the overhead in step 4 is $Q_1(n)$ time.

Step 5 is relatively straightforward after the previous steps. For each face, we collect the new chains and arcs that bound regions where the face is visible and we concatenate them to obtain the new cycles. This can be accomplished by sorting the endpoints of the chains and arcs to see which chains or arcs have to be connected to each other. The concatenation is easy because the chains are represented as concatenable queues. Thus step 5 takes time $O(k \log n)$.

After the above discussion the following lemma is straightforward.

Lemma 16.3 *The visibility map of a c-oriented set of polyhedra can be maintained in*

$$O(\, k \log n + Q_1(n) + (k+1)Q_4(k_S) + Q_5(k_S) + \sum_{1 \leqslant i \leqslant 3} U_i(n) + k \sum_{4 \leqslant i \leqslant 5} U_i(k_S) \,)$$

time per insertion, where k is the number of changes in the visibility map and k_S is the total complexity of the map.

16.3   Deletions

We now turn our attention to the deletion of a polyhedron $\mathcal{P}$. Because the new regions that appear after the deletion of $\mathcal{P}$ have nothing to do with $\mathcal{P}$ itself—they were just hidden by $\mathcal{P}$—deletions are harder than insertions. Indeed, one can obtain an output-sensitive solution to the static hidden surface problem by first inserting a large face that hides the whole scene, then insert the polyhedra of the scene, and finally remove the large face. Now the deletion algorithm has to discover the whole visibility map. If we do not insert the large face before inserting the other faces, then this incremental approach fails: the absence of a depth order makes that we cannot be sure whether a feature of the visibility map that we discover in some intermediate stadium of the algorithm will be visible at the end. Hence, an incremental algorithm that uses insertions only would not be output-sensitive.

Next we describe the basic algorithm for the deletion of a face f of $\mathcal{P}$. Recall that the visibility map can be viewed as a planar graph located on the viewing plane. (Although our data structures store the sets A and N of arcs and nodes that have been lifted to the background face of the corresponding regions.) When we use the term *component* in the sequel, we mean a component of this graph.

Algorithm 16.4
Input: A polyhedron $\mathcal{P} \in S$.
Output: The new visibility map $\mathcal{M}(S - \{\mathcal{P}\})$.
1.   Delete the faces and vertices of $\mathcal{P}$ from $\mathcal{D}_1$ and $\mathcal{D}_3$.
2.   **for** each outer boundary cycle $\mathcal{B}_i$ that bounds a region $\mathcal{R}_i$ where $\mathcal{P}$ is visible
3.   **do** { compute the new visibility map 'inside' $\mathcal{R}_i$ }
4.   Compute the new arcs of the component to which $\mathcal{B}_i$ belongs.
5.   Compute the arcs of the new components that 'float' inside $\mathcal{B}_i$.
6.   Remove the arcs that are defined by f from the current cycles.
7.   Assemble the new cycles using the arcs computed in step 4–5.
8.   Update the data structures that store R, A and N.
9.   Delete the curtains hanging from edges of $\mathcal{P}$ from $\mathcal{D}_2$.

Next we discuss the various steps of the algorithm in more detail.

In step 1 we already update the data structures $\mathcal{D}_1$ and $\mathcal{D}_3$ in time $U_1(n) + U_3(n)$. This is necessary because these structures will be used in step 2–5 to compute the new parts of the visibility map. If we would not do this at this point, then the old visibility map would be rediscovered in step 2–5. However, for reasons to become clear later we do not remove the curtains hanging from the edges of $\mathcal{P}$ from $\mathcal{D}_2$ yet.

The heart of the algorithm is, of course, in steps 3–5 where we discover the new arcs inside a region $\mathcal{R}_i$. See Figure 16.1, where the arcs of $\mathcal{B}_i$ are drawn fat, the new arcs that are attached to $\mathcal{B}_i$ are drawn normal, and the floating components are dotted. To compute the new parts of the visibility map we use the same strategy as

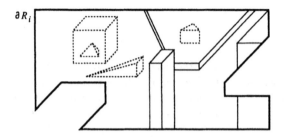

Figure 16.1: The new arcs that have to be discovered.

in Chapters 14 and 15: we try to find a node on every new component, and then we trace the arcs of the map using Lemma 14.1.

First, let us consider the arcs that are attached to $\mathcal{B}_i$. Observe that these new arcs will be in the same component of the visibility map as $\mathcal{B}_i$. When tracing this component, we have to restrict ourselves to what happens inside $\mathcal{R}_i$. Fortunately, the curtains that are hanging from the edges that define the arcs on the boundary of $\mathcal{R}_i$ ensure that the rays that we shoot cannot leave $\mathcal{R}_i$. This is the reason that we did not yet delete the curtains hanging from the edges of $\mathcal{P}$ from $\mathcal{D}_2$. The next problem to address is the computation of a starting node for this component. But this is easy, since $\mathcal{B}_i$ is part of the component. Hence, we can take any node on $\mathcal{B}_i$ as a starting node for the tracing process. This way we also rediscover $\mathcal{B}_i$, but we are allowed to spend this extra time because the background face of $\mathcal{R}_i$ has been deleted and, hence, the arcs of $\mathcal{B}_i$ have to be relocated (and possibly split).

Discovering the floating components is the most difficult task of the deletion algorithm. Notice that the arcs that are attached to $\mathcal{B}_i$ subdivide $\mathcal{R}_i$ into a number of new regions $\mathcal{R}_{i,j}$. There can be components floating inside each new region, and the problem is to find a node on each component. To this end we prove a lemma which characterizes the leftmost node—that is, the node with smallest x-coordinate—of all components floating inside a region $\mathcal{R}_{i,j}$. After having found this leftmost node, we trace its component; then we find the leftmost node of the remaining components and trace its component, et cetera, until all components have been computed.

Lemma 16.5 *The leftmost node of all components floating inside a region $\mathcal{R}_{i,j}$ is the projection of the leftmost vertex in V that is above $bf(\mathcal{R}_{i,j})$ and projects onto $\mathcal{R}_{i,j}$.*

Proof: It is readily seen that for non-intersecting polyhedra the leftmost node of a component can only be the projection of a visible vertex. It remains to show that the leftmost vertex v above $bf(\mathcal{R}_{i,j})$ is indeed visible. Assume for a contradiction that v is hidden by some face f. If ∂f does not intersect $\partial \mathcal{R}_{i,j}$ to the left of v, then one of the vertices of f is to the left of v and above $bf(\mathcal{R}_{i,j})$, contradicting the definition of v. On the other hand, if ∂f intersects $\partial \mathcal{R}_{i,j}$ to the left of v then we also get a contradiction, as follows. Because the polyhedra—and, in particular, f and $bf(\mathcal{R}_{i,j})$—do not intersect, the intersection of $\partial \overline{f}$ and $\partial \mathcal{R}_{i,j}$ must be visible. But then v cannot project onto $\mathcal{R}_{i,j}$. □

How do we find this leftmost vertex? Obviously, the data structure $\mathcal{D}_3$ was designed for this task. Unfortunately, we can only query with a face of constant complexity, and $\mathcal{R}_{i,j}$ could be a very complex polygonal region. Hence, we must decompose $\mathcal{R}_{i,j}$ into manageable pieces. We can decompose $\mathcal{R}_{i,j}$ into a number of quadrilaterals by constructing its vertical adjacency map, that is, by drawing segments which are parallel to the y–axis from every node on $\partial \mathcal{R}_{i,j}$ to the opposite arc on $\partial \mathcal{R}_{i,j}$. This vertical decomposition of $\mathcal{R}_{i,j}$ can be computed in linear time [17]. Since $\mathcal{R}_{i,j}$ is a new region, we are allowed to spend this amount of time. There is one problem with this approach, however. After we have decomposed $\mathcal{R}_{i_j}$ into quadrilaterals, we find the leftmost node ν of a component using Lemma 16.5. Starting from this node we then trace the component that is attached to ν. When the computation of this component has been finished, we would like to find the leftmost vertex of the remaining components floating inside $\mathcal{R}_{i,j}$, again using Lemma 16.5. The problem is that the component that we just discovered forms a hole in $\mathcal{R}_{i,j}$. Hence, the old quadrilaterals cannot be used, and we must compute the vertical decomposition anew. Clearly, we cannot afford to compute a new vertical decomposition every time we discover a new—and possibly very small—component. To solve this problem, we do not compute the complete vertical decomposition before we start querying. Instead, we compute the quadrilaterals during a sweep with a vertical segment s. The sweep segment s moves from left to right. It halts at every node of $\partial \mathcal{R}_{i,j}$ and also at the leftmost node of every component floating inside $\mathcal{R}_{i,j}$. When we stop at the leftmost node of a component, then we first trace that component using the technique of Chapter 14. This new component forms a hole inside $\mathcal{R}_{i,j}$. Thus the outer boundary of the new component is part of the boundary of $\mathcal{R}_{i,j}$, and the sweep segment must halt at the nodes on this boundary as well. Moreover, the sweep segment is split into two segments at the leftmost node of a component. One of the pieces sweeps the part above the hole, and the other piece sweeps the part below the hole. When both pieces reach the rightmost node of the component, then they merge again into one. Furthermore, a new sweep segment emerges from every 'local minimum' of the region which we are sweeping. In other words, a new sweep segment starts at every node on the boundary of $\mathcal{R}_{i,j}$ both of whose incident arcs lie to the right, such as nodes ξ_1 and ξ_2 in Figure 16.2. This is similar to the topological sweep used by Mulmuley in [88]. In order to advance the sweep segment s, we must compute the leftmost event point to the right of s. Suppose that the current sweep segment is incident to a node ν. Let a be the arc on the opposite side of ν that is touched by s, and let a' be the arc that is incident to and to the right of ν. See Figure 16.2. We define a quadrilateral $quad(\nu)$ as follows. The left edge of $quad(\nu)$ is s, and the top and bottom edges of $quad(\nu)$ are parts of a and a'. Finally, the right edge of $quad(\nu)$ is a vertical segment that contains a node μ

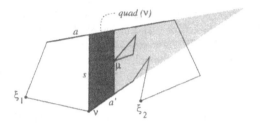

Figure 16.2: The computation of $quad(\nu)$.

defined as follows. Consider the (possibly unbounded) triangle to the right of s, which is determined by s and the half-lines through a and a'. In Figure 16.2 this triangle is depicted shaded. Node μ is the leftmost node of $\partial \mathcal{R}_{i,j}$ (including the nodes on the boundaries of its holes) inside this triangle. Observe that μ can be an endpoint of a or a'. Now the next event point where the sweep segment s must halt is either μ, or it is the leftmost node of a floating component inside $quad(\nu)$. To find node μ we can use $\mathcal{D}_5$. The leftmost node inside $quad(\nu)$ of a floating component can be found by querying with $quad(\nu)$—actually, we have to lift $quad(\nu)$ to $bf(\mathcal{R}_{i,j})$—according to Lemma 16.5, using $\mathcal{D}_3$.

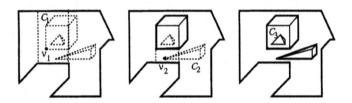

Figure 16.3: Computing the floating components.

So now we know how to sweep the region $\mathcal{R}_{i,j}$. When the sweep finishes, we have discovered all components floating in $\mathcal{R}_{i,j}$. However, there can be other components floating in regions of these new components. Hence, we have to treat these new regions recursively. The process is illustrated in Figure 16.3. The components that have not been found yet and the quadrilaterals with which we search are dotted, and the current boundary of $\mathcal{R}_{i,j}$—that is, the part that already has been discovered—is bold. First, a query is performed with the leftmost quadrilateral, but no node is found. Vertex ν_1 is found when searching with the next quadrilateral and its component $\mathcal{C}_1$ is computed. Then we discover ν_2 and $\mathcal{C}_2$. All the other searches fail, so we recurse in the regions of the new components. This way $\mathcal{C}_3$ is found. We recurse in the regions of $\mathcal{C}_3$ but we do not find any components and so we are finally ready.

The following lemma readily follows from the above discussion.

Lemma 16.6 *The visibility map of a set of polyhedra can be maintained in*

$$O\left((k+1)\left(\sum_{1 \leqslant i \leqslant 3} Q_i(n) + Q_5(k_S) \right) + \sum_{1 \leqslant i \leqslant 3} U_i(n) + k \sum_{4 \leqslant i \leqslant 5} U_i(k_S) \right)$$

time per deletion.

16.4 Axis-Parallel Polyhedra

To apply the result of the previous section we must implement the data structures $\mathcal{D}_1$–$\mathcal{D}_5$. We need not spend many words on the ray shooting structures, since we can apply results from Part B. However, two remarks are in place: First, the number of possible directions of the query rays for the structures $\mathcal{D}_2$ and $\mathcal{D}_4$ is constant—like in the static case—since we only shoot along edges or their projections. Hence, we can apply the same trick, which is to build separate structures for each possible direction. Second, the number of different orientations of the curtains hanging from the lifted arcs is also bounded. Using Theorems 5.5 and 6.17, we obtain the following results.

- $Q_1(m) = O(\log^2 m)$, $U_1(m) = O(\log^2 m \log \log m)$ and $M_1(m) = O(m \log^2 m)$.

- $Q_2(m) = Q_4(m) = O(\log^2 m)$, $U_2(m) = U_4(m) = O(\log^2 m)$ and $M_2(m) = M_4(m) = O(m \log m)$.

It remains to develop a structure for the following range query problem: Report the leftmost point—or all points—of a given set of points in $\mathbb{E}^3$ below (or above) an axis-parallel query face. (The leftmost point is the point with minimum y–coordinate.) For the axis-parallel case, this is quite easy. We briefly sketch a structure with $O(\log^2 n)$ update and query time, which uses $O(n \log n)$ storage. Using the decomposition scheme of Section 4.3.1, we can decompose any axis-parallel query face into a number of axis-parallel rectangles. So let us consider query rectangles with edges parallel to the x- and y-axis. Now we can use a structure due to Bern [14], which is a one-dimensional range tree on the y-coordinates of the points. At each node ν in this tree, we have an associated structure on the points whose y-coordinate is stored in the subtree rooted at this node. This associated structure is a priority search tree on the x- and z-coordinates of the points. We thus obtain the following performance for the range searching structures $\mathcal{D}_3$ and $\mathcal{D}_5$.

- $Q_3(m) = Q_5(m) = O(\log^2 m)$, $U_3(m) = U_5(m) = O(\log^2 m)$ and $M_3(m) = M_5(m) = O(m \log m)$.

Putting it all together, we obtain the following theorem.

Theorem 16.7 *The visibility map of a set of axis-parallel polyhedra with n vertices in total can be maintained in $O(\log^2 n \log \log n + k \log^2 n)$ amortized time per insertion or deletion of a polyhedron of constant complexity, where k is the number of changes in the view. The method uses $O((n+k_S) \log^2 n)$ storage, where k_S is the combinatorial complexity of the visibility map.*

16.5   *c*-Oriented Polyhedra

We now turn our attention to *c*-oriented polyhedra. Note that the number of possible directions of the query rays for the structures $\mathcal{D}_2$ and $\mathcal{D}_4$ is bounded by $O(c^3)$, as in the static case. The number of different orientations of the curtains hanging from the lifted arcs is also $O(c^3)$. By treating the different orientation separately and applying Theorems 5.9 and 6.20, we obtain the following results.

- $Q_1(m) = O(c^2 \log^2 m)$, $U_1(m) = O(\log^3 m)$ and $M_1(m) = O(m \log^2 m)$.

- $Q_2(m) = O(c \log^2 m)$, $U_2(m) = O(c^3 \log^2 m)$ and $M_2(m) = O(c^3 m \log m)$.

- $Q_4(m) = O(c^3 \log^2 m)$, $U_4(m) = O(c^3 \log^2 m)$ and $M_4(m) = O(c^3 m \log m)$.

(The update time of $\mathcal{D}_1$ worse than in Theorem 5.9, but now it is worst-case instead of amortized. This is advantageous, because the update time for the structures $\mathcal{D}_3$ and $\mathcal{D}_5$ discussed below is already $O(c^4 \log^3 m)$.)

Next, we develop a structure for the above mentioned range queries with *c*-oriented faces: Report the leftmost point—or, all points—of a given set of points in $\mathbb{E}^3$ below (or above) a *c*-oriented query face. Let us start by considering the following planar range searching problem: Preprocess a set P of points in the plane such that the leftmost point—or, all points—of P inside a *c*-oriented query polygon $\mathcal{P}$ can be reported efficiently. (The leftmost point is the point with minimum y–coordinate.) This problem has also been studied by Güting [67]. He obtains a data structure of size $O(c^2 n \log^2 n)$ such that queries take time $O(\log^2 n)$ and updates take time $O(c^2 \log^2 n)$. We reduce the amount of storage to $O(c^2 n \log n)$ without affecting the query and update time.

Recall from Section 4.3.2 that any *c*-oriented polygon $\mathcal{P}$ can be decomposed into a number of quadrilaterals, such that each resulting quadrilateral has two edges which are parallel to the y-axis, and two other edges—called the top and the bottom edge—which are *c*-oriented. We say that two quadrilateral are of the same type if and only if their top sides are parallel and their bottom sides are parallel. Since the query polygons are *c*-oriented, only $O(c^2)$ different types are possible for the resulting quadrilaterals. We build a separate structure for each type. With each quadrilateral resulting from the decomposition of $\mathcal{P}$, we search in the structure of the corresponding type; the answer to the query is easily computed from the constant number of subanswers that we find.

Consider a quadrilateral of a fixed type. This quadrilateral has its left and right side parallel to the y–axis, its top side has a fixed direction and its bottom side has a fixed direction. If we disregard the top side, then we are searching for the leftmost point in a half-infinite vertical slab and the problem is easily solved in $O(n)$ storage with $O(\log n)$ query and update time using a priority search tree [82]. Note that the priority search tree can also be used to report all points in the slab. Taking the top side into account means adding a range restriction to the searching problem. This can be done at the cost of an extra factor of $O(\log n)$ in the storage requirement and the query and update time, using the general techniques of Willard and Lueker [123].

Since we have used only standard structures and techniques, we leave the details to the reader. We obtain the following lemma.

Lemma 16.8 *Let P be a set of n points in the plane. The leftmost point, or all k points, of P inside a c-oriented polygon $\mathcal{P}$ can be found in $O(\log^2 n)$ time, respectively in $O(\log^2 n + k)$ time, with a structure that uses $O(c^2 n \log n)$ storage. The structure is dynamic and has an update time of $O(c^2 \log^2 n)$.*

Using this planar structure, we immediately obtain a data structure for the three-dimensional problem we set out to solve.

Corollary 16.9 *Let V be a set of n points in three-dimensional space. The leftmost point, or all k points, of V below (or above) a c-oriented face f can be found in $O(\log^3 n)$ time, respectively in $O(\log^3 n + k)$ time, with a structure that uses $O(c^2 n \log^2 n)$ storage. The structure is dynamic and has an update time of $O(c^2 \log^3 n)$.*

Proof: As before, we decompose a query face into quadrilaterals using the scheme of Section 4.3.2. We build a separate structure for each possible type of query quadrilateral. Such a quadrilateral has a fixed orientation, that is, it is parallel to a fixed plane. For a fixed type of quadrilaterals, the structure consists of the two-dimensional structure of Lemma 16.8 with another range restriction added to select the points that satisfy the restriction in the third dimension. Observe that the type of the quadrilaterals is fixed when we apply Lemma 16.8, so the dependency on c remains $O(c^2)$. □

This result is used to implement the data structures $\mathcal{D}_3$ and $\mathcal{D}_5$. These data structures are queried with c-oriented faces, but also with quadrilaterals that are the projection of a c-oriented quadrilateral onto a c-oriented face. This implies that the query objects are actually $O(c^2)$-oriented. We therefore have

- $Q_3(m) = Q_5(m) = O(\log^3 m)$, $U_3(m) = U_5(m) = O(c^4 \log^3 m)$ and $M_3(m) = M_5(m) = O(c^4 m \log^2 m)$.

Combining this and the above mentioned bounds on $\mathcal{D}_1$, $\mathcal{D}_2$ and $\mathcal{D}_4$ with Theorem 16.1, we obtain the following theorem.

Theorem 16.10 *The visibility map of a set of c-oriented polyhedra with n vertices in total can be maintained in time $O(c^4(k+1)\log^3 n)$ per insertion or deletion of a polyhedron of constant complexity, where k is the number of changes in the view. The method uses $O(c^4(n + k_S)\log^2 n)$ storage, where k_S is the combinatorial complexity of the visibility map.*

16.6   Arbitrary Polyhedra

The ray shooting data structures that we need in the case of arbitrary polyhedra can be taken from Part B. In particular, we use Theorem 5.13 to implement $\mathcal{D}_1$ and Theorem 7.21 for $\mathcal{D}_2$ and $\mathcal{D}_4$, where we take $m = n^{4/3}$. This way the query and update times of these structures is $O(n^{1/3+\epsilon})$. It remains to devise a dynamic data structure that stores a set P of points such that the leftmost point above (or below) a query face can be reported efficiently. We briefly sketch such a structure. Store the points of P in the leaves of a binary search tree $\mathcal{T}$ on x-coordinate. For a node ν in $\mathcal{T}$ let

$P(\nu)$ be the set of points stored in the subtree below ν. If we can decide at each node ν whether $P(\nu)$ contains a point above a query triangle t, then we can also find the leftmost point above t. (How this is done exactly is left as an exercise to the reader.) The query time will be a logarithmic factor worse than the time we need to decide at a node ν whether $P(\nu)$ contains a point above t. To make this decision we store $P(\nu)$ in an associate structure as described in Lemma 11.11, which has $O(n^{1/3+\varepsilon})$ query and update time.

This leads to the following theorem.

Theorem 16.11 *Let $\varepsilon > 0$. The visibility map of a set of polyhedra with n vertices in total can be maintained in amortized time $O((k+1)(n+k_S)^{1/3+\varepsilon})$ per insertion or deletion of a polyhedron of constant complexity, where k is the number of changes in the view. The method uses $O((n+k_S)^{4/3+\varepsilon})$ storage, where k_S is the combinatorial complexity of the visibility map.*

Chapter 17

Conclusions

In this part of the book we presented an algorithm for computing the visibility map of a set of polyhedra. The most important feature of our algorithm is that it is output-sensitive and that a depth order on the polyhedra need not exist. None of the other techniques developed so far combines these two features. Table 17.1 gives an overview of the running times of our algorithm in all the static cases we studied. Here n is the total number of vertices of all polyhedra, k is the combinatorial complexity of the map, and ε is a constant that can be chosen arbitrarily small. For axis-parallel and

	non-intersecting	intersecting
axis-parallel	$(n+k)\log n$	$(n+k)\log n(\log\log n)^2$
c-oriented	$c^3 n\log n + c^2 k\log n$	$c^4 n\log^2 n + c^2 k\log n(\log\log n)^2$
arbitrary	$n^{1+\varepsilon} + n^{2/3+\varepsilon}k^{2/3}$	$n^{4/3+\varepsilon} + n^{4/5+\varepsilon}k^{4/5}$

Table 17.1: Results on computing the visibility map of a set of polyhedra.

c-oriented polyhedra we also obtained other bounds, which are more efficient for some values of k and c. We refer the reader to Chapters 14 and 15 for details. We have also obtained some results on dynamically maintaining the visibility map of a set of non-intersecting polyhedra under insertions into and deletions from the set, for which we refer the reader to Chapter 16.

Let us discuss how good the results are that we obtained for computing the visibility map of a set of non-intersecting polyhedra. Both for axis-parallel and for c-oriented polyhedra (with constant c) our results approach the trivial $\Omega(n+k)$ lower bound up a logarithmic factor. For arbitrary polyhedra the running time seems to be far from optimal, although it is close to optimal for constant as well as for quadratic values of k. But in general a wide gap remains between the upper and the lower bound. The question is, however, whether the upper bound is to blame for this entirely. We think that it is not. Thus it might be wise to try and prove non-trivial lower bounds for the problem of computing visibility maps in an output-sensitive manner. One of the problems in proving lower bounds is the fact that the bound has two parameters, n and k.

Suppose, for example, that we reduce the problem to *Element Uniqueness*, which is defined as follows: Given a set $S = \{x_1, \ldots, x_n\}$ of n real numbers, decide if any two of them are equal. We can solve this problem—for which an $\Omega(n \log n)$ lower bound is known [104]—by computing the visibility map of the set $S' = \{(x_i, 0, i) : 1 \leqslant i \leqslant n\}$ of points[1] in three-dimensional space for the vertical viewing direction: all numbers in S are distinct if and only if all points in S' are visible. But what can we conclude from this reduction? An $\Omega(n \log n + k)$ lower bound, an $\Omega(n + k \log k)$ lower bound, or perhaps even an $\Omega((n+k) \log n)$ lower bound? To avoid these problems, we can define the following decision problem: Given a set S of non-intersecting polyhedra with n vertices in total, decide whether the number of nodes of the visibility map is less than n. The reduction from Element Uniqueness, which we described above, proves an $\Omega(n \log n)$ lower bound for this problem. Our algorithm for computing visibility maps answers this question in $O(n^{4/3+\varepsilon})$ time. We believe that this is quasi-optimal, that is, that $\Omega(n^{4/3})$ is a lower bound for the problem. Let us try to motivate this belief.

Consider *Hopcroft's problem* in two-dimensional space: Given a set of $n/2$ lines and a set of $n/2$ points in the plane, decide whether any line contains any point. This question can be answered by giving the points height 0 and the lines height $1, \ldots, n/2$, and computing the complexity of the visibility map of the scene as seen from $z = \infty$. The best known solution to this problem has been given by Chazelle [18]. It runs in time $O(n^{4/3} \log^{1/3} n)$. Another related problem is to count the number of intersections of a set of line segments in the plane, whose best known solution is also $O(n^{4/3} \log^{1/3} n)$ [18]. The exponent 4/3 does not pop up in all these bounds coincidentally: the data structures that support our algorithm for computing visibility maps, and Chazelle's algorithms for Hopcroft's problem and for the problem of counting segment intersections, are all based on the same geometric partitionings of space, namely cuttings and simplicial partitions. The bounds on these partionings are (asymptotically) optimal, see Sections 2.5 and 2.6. The question is whether the problems are so closely tied to these partitionings that $\Omega(n^{4/3})$ is a lower bound, or that a completely different approach exists that beats this bound. We expect the first to be true.

Up to now we have assumed that the objects in the scene are polyhedra. In practice, there are often other kinds of objects present—spheres, cylinders, and so on— but usually it is sufficient to approximate the surface patches that form these objects by a number of small polygons. It should be noted, however, that the applicability of our algorithm for non-intersecting objects is not restricted to polyhedral objects. Of course, if we want to apply our algorithm we need to be able to answer certain ray shooting queries. As an example, let us consider the case of discs in $\mathbb{E}^3$. What we need is a data structure for ray shooting from infinity into the viewing direction in the set of discs, and a structure for ray shooting in the set of 'curtains' hanging from the boundary of the discs. Recall that the shooting in curtains is done with rays along (projected) boundaries. In the case of polyhedra these boundaries consist of edges, but in the case of discs the boundaries are circles. This means that we are shooting 'along a circle'. If we can develop a data structure for such 'curved ray shooting queries' in curtains hanging from circles, then we can apply our method. We think that this type of curved ray shooting problems is worth studying. Note that our

---

[1]The construction is easily modified such that it results in a set of tetrahedra instead of points.

algorithm needs a visible vertex on each component of the visibility map; in the case of discs it suffices to define the leftmost point of each disc to be a vertex. This also indicates the problem when we study curved objects that are allowed to intersect: we can no longer guarantee that each component of the visibility map contains a visible vertex. It would be interesting to find a new output-sensitive method that is also able to handle intersecting curved objects.

References

[1] P.K. Agarwal, Ray Shooting and Other Applications of Spanning Trees with Low Stabbing Number, *Proc. 5th ACM Symp. on Computational Geometry*, 1989, pp. 315–325.

[2] P.K. Agarwal, Partitioning Arangements of Lines I: An Efficient Deterministic Algorithm, *Discr. & Computational Geometry* 5 (1990) pp. 449–483.

[3] P.K. Agarwal and J. Matoušek, Ray Shooting and Parametric Search, *Proc. 24th ACM Symp. on Theory of Computing*, 1992, pp. 517–526.

[4] P.K. Agarwal and J. Matoušek, *On Range Searching with Semialgebraic Sets*, Tech. Rep. CS-1992-13, Dept. of Computer Science, Duke University, 1992.

[5] P.K. Agarwal, D. Eppstein and J. Matoušek, Dynamic Half-Space Reporting, Geometric Optimization and Minimum Spanning Trees, *Proc. 33rd IEEE Symp. on Foundations of Computer Science*, 1992, pp. 80–89.

[6] P.K. Agarwal and M. Sharir, Applications of a New Space Partitioning Technique, *Discr. & Computational Geometry* 9 (1993) pp. 11–38.

[7] P.K. Agarwal, M. van Kreveld and M. Overmars, Intersection Queries for Curved Objects, *Proc. 7th ACM Symp. on Computational Geometry*, 1991, pp. 41–50.

[8] A.V. Aho, J.E. Hopcroft and J.D. Ullman, *The Design and Analysis of Computer Algorithms*, Addison-Wesley, Reading, Mass., 1974.

[9] H. Alt, R. Fleischer, M. Kaufmann, K. Mehlhorn, S. Näher, S. Schirra and C. Uhrig, Approximate Motion Planning and the Complexity of the Boundary of the Union of Simple Geometric Figures, *Proc. 6th ACM Symp. on Computational Geometry*, 1990, pp. 281–289.

[10] B. Aronov and M. Sharir, On the Zone of a Surface in a Hyperplane Arrangement, *Proc. Workshop on Algorithms and Data Structures*, Lecture Notes in Computer Science 519, 1991, pp. 13–19.

[11] T. Asano, T. Asano, L. Guibas, J. Hershberger and H. Imai, Visibility of Disjoint Polygons, *Algorithmica* 1 (1986), pp. 49–63.

[12] R. Bar-Yehuda and R. Grinwald, Triangulating Polygons with Holes, *Proc. 2nd Canadian Conference on Computational Geometry*, 1990, pp. 112–115.

[13] J.L. Bentley, *Algorithms for Klee's Rectangle Problems*, unpublished note, Dept. of Computer Science, Carnegie-Mellon University, 1977.

[14] M. Bern, Hidden Surface Removal for Rectangles, *J. Comp. Syst. Sciences* **40** (1990), pp. 49–69.

[15] B. Chazelle, A Theorem on Polygon Cutting with Applications, *Proc. 23rd IEEE Symp. on Foundations of Computer Science*, 1982, pp. 339–349.

[16] B. Chazelle, Lower Bounds on the Complexity of Polytope Range Searching, *J. Amer. Math. Soc.* **2** (1989), pp. 637–666.

[17] B. Chazelle, Triangulating a Simple Polygon in Linear Time, *Discr. & Computational Geometry* **6** (1991) pp. 485–524.

[18] B. Chazelle, An Optimal Convex Hull Algorithm and New Results on Cuttings, *Proc. 32nd IEEE Symp. on Foundations of Computer Science*, 1991, pp. 29–38.

[19] B. Chazelle, Computational Geometry for the Gourmet: Old Fare and New Dishes, *Proc. 18th Int. Coll. on Automata, Languages and Programming*, Lecture Notes in Computer Science 510, 1991, pp. 686–696.

[20] B. Chazelle, H. Edelsbrunner, M. Gringi, L.J. Guibas, M. Sharir and J. Snoeyink, Ray Shooting in Polygons Using Geodesic Triangulations, *Proc. 18th Int. Coll. on Automata, Languages and Programming*, Lecture Notes in Computer Science 510, 1991, pp. 661–673.

[21] B. Chazelle, H. Edelsbrunner, L. Guibas, R. Pollack, R. Seidel, M. Sharir and J. Snoeyink, Counting and Cutting Cycles of Lines and Rods in Space, *Proc. 31st IEEE Symp. on Foundations of Computer Science*, 1990, pp. 242–251.

[22] B. Chazelle, H. Edelsbrunner, L.J. Guibas and M. Sharir, Lines in Space — Combinatorics, Algorithms and Applications, *Proc. 21st ACM Symp. on Theory of Computing*, 1989, pp. 382–393.

[23] B. Chazelle and J. Friedman, A Deterministic View of Random Sampling and its Use in Computational Geometry, *Combinatorica* **10** (1990), pp. 229–249.

[24] B. Chazelle and J. Friedman, *Point Location among Hyperplanes and Unidirectional Ray-Shooting*, Tech. Rep. CS-TR-333-91, Princeton University, 1991.

[25] B. Chazelle and L.J. Guibas, Fractional Cascading I: A Data Structuring Technique, *Algorithmica* **1** (1986), pp. 133–162.

[26] B. Chazelle and L.J. Guibas, Fractional Cascading II: Applications, *Algorithmica* **1** (1986), pp. 163–191.

[27] B. Chazelle and L. Guibas, Visibility and Intersection Problems in Plane Geometry, *Discr. & Computational Geometry* **4** (1989) pp. 551–581.

[28] B. Chazelle, M. Sharir and E. Welzl, Quasi-Optimal Upper Bounds for Simplex Range Searching and New Zone Theorems, *Proc. 6th ACM Symp. on Computational Geometry*, 1990, pp. 23–33.

[29] B. Chazelle and E. Welzl, Quasi-Optimal Range Searching in Spaces of Finite VC-Dimension, *Discr. & Computational Geometry* **4** (1989) pp. 467–489.

[30] S.W. Cheng, *Dynamic Hidden Line Elimination*, manuscript, 1991.

[31] S.W. Cheng and R. Janardan, New Results on Dynamic Planar Point Location, *Proc. 31st IEEE Symp. on Foundations of Computer Science*, 1990, pp. 96–105.

[32] S.W. Cheng and R. Janardan, Space-Efficient Ray-Shooting and Intersection Searching: Algorithms, Dynamization and Applications, *Proc. 2nd ACM-SIAM Symp. on Discrete Algorithms*, 1991, pp. 7–16.

[33] F. Chin and C.A. Wang, Optimal Algorithms for the Intersection and the Minimum Distance Problems Between Planar Polygons, *IEEE Trans. on Computers* **C-32** (1983), pp. 1203–1207.

[34] K. Clarkson, New Applications of Random Sampling in Computational Geometry, *Discr. & Computational Geometry* **2** (1987), pp. 195–222.

[35] K. Clarkson, A Randomized Algorithm for Closest-Point Queries, *SIAM J. Comput.* **17** (1988), pp. 830–847.

[36] R. Cole and M. Sharir, Visibility Problems for Polyhedral Terrains, *J. Symbolic Computation* **7** (1989), pp. 11–30.

[37] T.H. Cormen, C.E. Leiserson and R.L. Rivest, *Introduction to Algorithms*, MIT Press, Cambridge, 1990.

[38] M. de Berg, Translating Polygons with Applications to Hidden Surface Removal, *Proc. 2nd Scandinavian Workshop on Algorithm Theory*, 1990, Lecture Notes in Computer Science 447, 1990, pp. 60–70.

[39] M. de Berg, Dynamic Output-Sensitive Hidden Surface Removal for c–Oriented Polyhedra, *Computational Geometry: Theory and Applications* **2** (1992), pp. 119–140.

[40] M. de Berg, H. Everett and H. Wagener, *Translation Queries for Sets of Polygons*, Tech. Rep. RUU-CS-91-30, Dept. of Computer Science, Utrecht University, 1991.

[41] M. de Berg, D. Halperin, M.H. Overmars, J. Snoeyink and M. van Kreveld, Efficient Ray Shooting and Hidden Surface Removal, *Proc. 7th ACM Symp. on Computational Geometry*, 1991, pp. 21–30. To appear in *Algorithmica*.

[42] M. de Berg and M.H. Overmars, Hidden Surface Removal for Axis-Parallel Polyhedra, *Proc. 31st IEEE Symp. on Foundations of Computer Science*, 1990, pp. 252–261.

[43] M. de Berg and M.H. Overmars, Hidden Surface Removal for c–Oriented Polyhedra, *Computational Geometry: Theory and Applications* **1** (1992), pp. 247–268.

[44] M. de Berg, M.H. Overmars and O. Schwarzkopf, *Computing and Verifying Depth Orders*, Proc. *8th ACM Symp. on Computational Geometry*, 1992, pp. 138–145.

[45] M. de Berg, M. van Kreveld and J. Snoeyink, *Two- and Three-Dimensional Point Location in Rectangular Subdivisions*, Tech. Rep. RUU-CS-91-29, Dept. of Computer Science, Utrecht University, 1991.

[46] F. Dehne and J.-R. Sack, Separability of Sets of Polygons, *Proc. 12th International Workshop on Graph-Theoretic Concepts in Computer Science*, 1986, pp. 237–251.

[47] F. Dévai, Quadratic Bounds for Hidden Line Elimination, *Proc. 2nd ACM Symp. on Computational Geometry*, 1986, pp. 269–275.

[48] D.P. Dobkin and D. Kirkpatrick, A Linear Algorithm for Determining the Separation of Convex Polyhedra, *J. Algorithms* **6** (1985), pp. 381–392.

[49] H. Edelsbrunner, Computing the Extreme Distances between Two Convex Polygons, *J. of Algorithms* **6** (1985), pp. 213–224.

[50] H. Edelsbrunner, *Algorithms in Combinatorial Geometry*, Springer-Verlag, Berlin, 1987.

[51] H. Edelsbrunner, L.J. Guibas and M. Sharir, The Upper Envelope of Piecewise Linear Functions: Algorithms and Applications, *Discr. & Computational Geometry* **4** (1989), pp. 311–336.

[52] H. Edelsbrunner and H.A. Maurer, On the Intersection of Orthogonal Objects, *Inf. Proc. Lett.* **13** (1981), pp. 177–181.

[53] H. Edelsbrunner, J. Pach, J.T. Schwartz and M. Sharir, On the Lower Envelope of Bivariate Functions and its Applications, *Proc. 28th IEEE Symp. on Foundations of Computer Science*, 1987, pp. 27–37.

[54] H. Edelsbrunner, A.D. Robison and X.J. Shen, Covering Convex Sets with Non-Overlapping Polygons, *Discrete Mathematics* **81** (1990), pp. 153–164.

[55] H. Edelsbrunner and E. Welzl, Halfplanar Range Search in Linear Space and $O(n^{0.695})$ Query Time, *Inf. Proc. Lett.* **23** (1986), pp. 289–293.

[56] P. Egyed, Hidden Surface Removal in Polyhedral–Cross–Sections, *The Visual Computer* **3** (1988), pp. 329–343.

[57] H.A. El Gindy and G.T. Toussaint, Efficient Algorithms for Inserting and Deleting Edges from Triangulations, *Proc. Int. Conf. on Foundations of Data Organization*, 1985.

[58] J.D. Foley, A. van Dam, S.K. Feiner and J.F. Hughes, *Computer Graphics: Principles and Practice*, Addison-Wesley, Reading, Mass., 1990.

[59] H. Fuchs, Z. Kedem and B. Naylor, On Visible Surface Generation by A Priori Tree Structures, *Computer Graphics, Proc. SIGGRAPH '80*, 1980, pp. 124-133.

[60] H.N. Gabow, J.L. Bentley and R.E. Tarjan, Scaling and Related Techniques for Geometry Problems, *Proc. 16th ACM Symp. on Theory of Computing*, 1984, pp. 135–143.

[61] A.S. Glassner (Ed.), *An Introduction to Ray Tracing*, Academic Press, London, 1989.

[62] M.T. Goodrich, A Polygonal Approach to Hidden Line Elimination, *Proc. 25th Allerton Conf. on Communication, Control and Computing*, 1987, pp. 849–858.

[63] M.T. Goodrich, M.J. Atallah and M.H. Overmars, An Input-Size/Output-Size Trade-Off in the Time-Complexity of Rectilinear Hidden Surface Removal, *Proc. 17th Int. Coll. on Automata, Languages and Programming*, Lecture Notes in Computer Science 443, 1990, pp. 689–702.

[64] L.J. Guibas, M. Overmars and M. Sharir, *Ray Shooting, Implicit Point Location and Related Queries in Arrangements of Segments*, Tech. Rep. No. 443, Courant Institute of Math. Sciences, New York University, 1989.

[65] L.J. Guibas and F.F. Yao, On Translating a Set of Rectangles, in: F.P. Preparata (Ed.), *Advances in Computing Research, Vol. I: Computational Geometry*, JAI Press Inc., 1983, pp. 61–77.

[66] R.H. Güting, Stabbing c–Oriented Polygons, *Inf. Proc. Lett.* **16** (1983), pp. 35–40.

[67] R.H. Güting, *Conquering Contours: Efficient Algorithms for Computational Geometry*, Ph.D. Thesis, Universität Dortmund, 1983.

[68] R.H. Güting and T. Ottmann, *New Algorithms for Special Cases of the Hidden Line Elimination Problem*, Forschungsbericht 184, Abteilung Informatik, Universität Dortmund, 1984.

[69] R.H. Güting and T. Ottmann, New Algorithms for Special Cases of the Hidden Line Elimination Problem, *Comp. Vision, Graphics and Image Processing* **40** (1987), pp. 188–204.

[70] D. Haussler and E. Welzl, Epsilon-Nets and Simplex Range Queries, *Discr. & Computational Geometry* **2** (1987), pp. 127–151.

[71] W.V.D. Hodge and D. Pedoe, *Methods of Algebraic Geometry*, Cambridge University Press, 1952.

[72] M.J. Katz, M.H. Overmars and M. Sharir, Efficient Hidden Surface Removal for Objects with Small Union Size, *Proc. 7th ACM Symp. on Computational Geometry*, 1991, pp. 31–40.

[73] D.G. Kirkpatrick, Optimal Search in Planar Subdivisions, *SIAM J. Comput.* **12** (1983), pp. 28–35.

[74] D.E. Knuth, *The Art of Computer Programming I: Fundamental Algorithms*, Addison-Wesley, Reading, Mass., 1968.

[75] D.T. Lee and F.P. Preparata, Euclidean Shortest Paths in the Presence of Rectilinear Barriers, *Networks* **14** (1984), pp. 393–410.

[76] W. Lipsky and F.P. Preparata, Segments, Rectangles, Contours, *J. Algorithms* **2** (1981), pp. 63–76.

[77] J. Matoušek, Construction of ε-Nets, *Discr. & Computational Geometry* **5** (1990), pp. 427–448.

[78] J. Matoušek, Cutting Hyperplane Arrangements, *Proc. 6th ACM Symp. on Computational Geometry*, 1990, pp. 1–9.

[79] J. Matoušek, Approximations and Optimal Geometric Divide-and-Conquer, *Proc. 23rd ACM Symp. on Theory of Computing*, 1991, pp. 506–511.

[80] J. Matoušek, Efficient Partition Trees, *Proc. 7th ACM Symp. on Computational Geometry*, 1991, pp. 1–9.

[81] J. Matoušek, Range Searching with Efficient Hierarchical Cuttings, *Proc. 8th ACM Symp. on Computational Geometry*, 1992, pp. 276–285.

[82] E.M. McCreight, Priority Search Trees, *SIAM J. Comput.* **14** (1985), pp. 257–276.

[83] M. McKenna, Worst-Case Optimal Hidden Surface Removal, *ACM Trans. Graphics* **6** (1987) pp. 19–28.

[84] M. McKenna and J. O'Rourke, Arrangements of Lines in 3-Space: A Data Structure with Applications, *Proc. 4th ACM Symp. on Computational Geometry*, 1988, pp. 371–380.

[85] K. Mehlhorn, *Data Structures and Algorithms 3: Multi-Dimensional Searching and Computational Geometry*, Springer-Verlag, Berlin, 1984.

[86] K. Mehlhorn and S. Näher, Dynamic Fractional Cascading, *Algorithmica* **5** (1990), pp. 215–241.

[87] K. Mulmuley, An Efficient Algorithm for Hidden Surface Removal, I, *Computer Graphics* **23** (1989), pp. 379–388.

[88] K. Mulmuley, On Obstructions in Relation to a Fixed Viewpoint, *Proc. 30th IEEE Symp. on Foundations of Computer Science*, 1989, pp. 592–597.

[89] O. Nurmi, *On Translating a Set of Objects in Two- and Three-Dimensional Space*, Bericht 141, Insitut für Angewandte Informatik und Formale Beschreibungsverfahren, Universität Karlsruhe, 1984.

[90] O. Nurmi, A Fast Line-Sweep Algorithm for Hidden Line Elimination, *BIT* **25** (1985) pp. 466–472.

[91] O. Nurmi, On Translating a Set of Objects in Two- and Three-Dimensional Space, *Computer Vision, Graphics and Image Processing* **36** (1986), pp. 42–52.

[92] D. Nussbaum and J.-R. Sack, Disassembling Two-Dimensional Composite Parts Via Translations, *Proc. Int. Conf. on Optimal Algorithms*, 1989.

[93] D. Nussbaum and J.-R. Sack, *Translation Separability of Polyhedra*, manuscript, presented at the 1st Canadian Conf. on Computational Geometry.

[94] T. Ottmann and P. Widmayer, On Translating a Set of Line Segments, *Computer Vision, Graphics and Image Processing* **24** (1983), pp. 382–389.

[95] M.H. Overmars, *The Design of Dynamic Data Structures*, Lecture Notes in Computer Science 156, Springer-Verlag, Berlin, 1983.

[96] M.H. Overmars, H. Schipper and M. Sharir, Storing Line Segments in Partition Trees, *BIT* **30** (1990), pp. 385–403.

[97] M.H. Overmars and M. Sharir, Output-Sensitive Hidden Surface Removal, *Proc. 30th IEEE Symp. on Foundations of Computer Science*, 1989, pp. 598–603.

[98] M.H. Overmars and M. Sharir, *An Improved Technique for Output-Sensitive Hidden Surface Removal*, Tech. Rep. RUU-CS-89-32, Dept. of Comp. Science, Utrecht University, 1989.

[99] J. Pach and M. Sharir, The Upper Envelope of Piecewise Linear Functions and the Boundary of a Region Enclosed by Convex Plates: Combinatorial Analysis, *Discr. & Computational Geometry* **4** (1989), pp. 291-309.

[100] M.S. Paterson and F.F. Yao, Efficient Binary Partitions for Hidden-Surface Removal and Solid Modelling, *Discr. & Computational Geometry* **5** (1990), pp. 485–503.

[101] M. Pellegrini, Stabbing and Ray Shooting in 3-Dimensional Space, *Proc. 6th ACM Symp. on Computational Geometry*, 1990, pp. 177–186.

[102] M. Pellegrini, New Results on Ray Shooting and Isotopy Classes of Lines in 3-Dimensional Space, *Proc. Workshop on Algorithms and Data Structures*, Lecture Notes in Computer Science 519, 1991, pp. 20–31.

[103] M. Pellegrini, On the Zone of a Co-Dimension p Surface in a Hyperplane Arrangement, *Proc. 3rd Canadian Conference on Computational Geometry*, 1991, pp. 233–238.

[104] F.P. Preparata and M.I. Shamos, *Computational Geometry, An Introduction*, Springer-Verlag, New York, 1985.

[105] F.P. Preparata and K.J. Supowit, Testing a Simple Polygon for Monotonicity, *Inf. Proc. Lett.* **12** (1981), pp.161–164.

[106] F.P. Preparata, J.S. Vitter and M. Yvinec, Computation of the Axial View of a Set of Isothetic Parallelepipeds, *ACM Trans. on Graphics* **9** (1990), pp. 278–300.

[107] F.P. Preparata, J.S. Vitter and M. Yvinec, Output-Sensitive Generation of the Perspective View of Isothetic Parallelepipeds, *Proc. 2nd Scandinavian Workshop on Algorithm Theory*, 1990, Lecture Notes in Computer Science 447, 1990, pp. 71–84.

[108] J. Reif and S. Sen, An Efficient Output-Sensitive Hidden Surface Removal Algorithm and its Parallelization, *Proc. 4th ACM Symp. on Computational Geometry*, 1988, pp. 193–200.

[109] J.-R. Sack and G.T. Toussaint, Translating Polygons in the Plane, *Proc. 2nd Annual Symp. on Theoretical Aspects of Computer Science*, 1985, pp. 310–321.

[110] A. Schmitt, Time and Space Bounds for Hidden Line and Hidden Surface Algorithms, *Eurographics '81*, pp. 43–56.

[111] A. Schmitt, H. Müller and W. Leister, Ray Tracing Algorithms — Theory and Practice, in: R.A. Earnshaw (Ed.), *Theoretical Foundations of Computer Graphics and CAD*, NATO ASI Series, Vol. F40, Springer-Verlag, 1988, pp. 997–1030.

[112] B. Scholten and J. van Leeuwen, Structured NC, *Proc. Workshop on Algorithms and Data Structures*, Lecture Notes in Computer Science 382, 1989, pp.487–499.

[113] J.T. Schwartz and M. Sharir, On the Piano Movers' Problem, I: The Case of a Two-Dimensional Rigid Polygonal Body Moving amidst Polygonal Barriers, *Comm. Pure Appl. Math.* **36** (1983), pp. 345–398.

[114] M.I. Shamos, Geometric Complexity, *Proc. 7th ACM Symp. on Theory of Computing*, 1975, pp. 224–233.

[115] M. Sharir and M.H. Overmars, A Simple Method for Output-Sensitive Hidden Surface Removal, *ACM Trans. on Graphics*, 1992, to appear.

[116] J. Stolfi, *Primitives for Computational Geometry*, Ph.D. Thesis, Computer Science Department, Stanford University, 1988.

[117] I.E. Sutherland, R.F. Sproull and R.A. Schumacker, A Characterization of Ten Hidden-Surface Algorithms, *Computing Surveys* **6** (1974) pp. 1–25.

[118] G.T. Toussaint, Movable Separability of Sets, in: G.T. Toussaint (Ed.), *Computational Geometry*, North Holland, 1985, pp. 335–376.

[119] G.T. Toussaint, On Separating Two Simple Polygons by a Single Translation, *Discr. & Computational Geometry* 4 (1989), pp. 265–278.

[120] A. Watt, *Fundamentals of Three-Dimensional Computer Graphics*, Addison-Wesley Publishing Company, 1989.

[121] R. Wenger, *Upper Bounds on Geometric Permutations for Convex Sets, Discr. & Computational Geometry* 5 (1990), pp. 27–33.

[122] D.E. Willard, Polygon Retrieval, *SIAM J. Comput.* 11 (1982), pp. 149–165.

[123] D.E. Willard and G.S. Lueker, Adding Range Restriction Capability to Dynamic Data Structures, *J. ACM* 32 (1985) pp. 597–617.

[124] F.F. Yao, On the Priority Approach to Hidden-Surface Algorithms, *Proc. 21st IEEE Symp. on Foundations of Computer Science*, 1980, pp. 301–307.

[125] F.F. Yao, Computational Geometry, In: J. van Leeuwen (Ed.), *Handbook of Theoretical Computer Science, Volume A: Algorithms and Complexity*, Elsevier Science Publishers, Amsterdam, 1990, pp. 343–389.

[126] C.K. Yap, How to Move a Chair through a Door, *IEEE Trans. on Robotics and Automation* **RA-3** (1987), pp. 173–181.

[127] C.K. Yap, A Geometric Consistency Theorem for a Symbolic Perturbation Scheme, *Proc. 4th ACM Symp. on Computational Geometry*, 1988, pp. 134–142.

Notation

Lower Case Roman

a	arc (in a graph, or in a visibility map)
c	number of orientations (as in c-oriented) or cell
d	fixed number denoting the dimension
e	edge (of a graph or a polyhedron)
f	face of a polytope
h	hyperplane
i, j	integers
k	output size of an algorithm
l	line
m, n	cardinality of a set
o	object
p	point, usually starting point of a ray or view point
q	point
r	size of sample R
s	simplex or segment
t	triangle or triangulation segment
v, w	vertices (of a graph, segment or polytope)

Upper Case Roman

A	set of arcs
D	set of directions
E	set of edges
F	set of faces
H	set of hyperplanes
L	set of lines
N	set of nodes
P	set of points
R	(random) sample, or set of regions
S	set
T	set of triangles
V	set of vertices

Script

$\mathcal{A}$	arrangement (sometimes triangulated)
$\mathcal{B}$	boundary cycle
$\mathcal{C}$	connected component
$\mathcal{CH}$	convex hull
$\mathcal{D}$	data structure
$\mathcal{E}$	upper envelope
$\mathcal{G}$	graph
$\mathcal{I}$	intersection sequence
$\mathcal{L}$	list
$\mathcal{M}$	visibility map
$\mathcal{Q}$	queue
$\mathcal{P}$	polygon or polyhedron
$\mathcal{R}$	region in a visibility map
$\mathcal{S}$	suffix
$\mathcal{SH}$	shadow
$\mathcal{T}$	tree
$\mathcal{V}$	vertical adjacency map

Lower Case Greek

β, γ	polygonal chains
δ, ε	positive real numbers, usually small
μ, ν, ξ	nodes in a tree, or in the visibility map
ρ	ray
π	Plücker point
ϖ	Plücker plane

Upper Case Greek

Ξ	cutting
Π	Plücker hypersurface
Σ	subdivision
Ψ	simplicial partitioning

Notation related to trees

$root(\mathcal{T})$	root of $\mathcal{T}$
$lchild(\nu)$	left child of node ν
$rchild(\nu)$	right child of node ν
$parent(\nu)$	parent of node ν
I_ν	interval corresponding to node ν in a segment tree

Functions related to running times, etc.

$\alpha(n)$	inverse of Ackermann's function
$D(n)$	deletion time
$I(n)$	insertion time
$M(n)$	storage requirement
$Q(n)$	query time
$T(n)$	preprocessing time
$U(n)$	update time

Miscellaneous

$|S|$	cardinality of the set S
$\mathbb{E}^d$	Euclidean space, superscripted with the dimension d
$\bar{o}$	projection of o, usually onto xy–plane
$\overline{pq}$	line segment with p and q as endpoints.
o^*	dual of o
$\prec_*$	transitive closure of the relation $\prec$
$l(o)$	line containing one-dimensional object o
p_x, p_y, p_z	coordinates of point p in $\mathbb{E}^3$
$p_1, \ldots, p_d$	coordinates of point p in $\mathbb{E}^d$
$int(o)$	interior of o
$ext(o)$	exterior of o
∂o	boundary of o
$\mathcal{CH}(\mathcal{P}|S)$	convex hull of $\mathcal{P}$ relative to S
$\mathcal{P}^\circ(S)$	$\mathcal{CH}(\mathcal{P}|S - \{\mathcal{P}\})$
$\Phi_\rho(S)$	first object in S hit by ρ
$bf(\mathcal{R})$	background face of region $\mathcal{R}$

Index

Springer-Verlag
and the Environment

We at Springer-Verlag firmly believe that an international science publisher has a special obligation to the environment, and our corporate policies consistently reflect this conviction.

We also expect our business partners – paper mills, printers, packaging manufacturers, etc. – to commit themselves to using environmentally friendly materials and production processes.

The paper in this book is made from low- or no-chlorine pulp and is acid free, in conformance with international standards for paper permanency.

Printing: Weihert-Druck GmbH, Darmstadt
Binding: Buchbinderei Schäffer, Grünstadt

Lecture Notes in Computer Science

For information about Vols. 1–620
please contact your bookseller or Springer-Verlag

Vol. 661: S. J. Hanson, W. Remmele, R. L. Rivest (Eds.), Machine Learning: From Theory to Applications. VIII, 271 pages. 1993.

Vol. 662: M. Nitzberg, D. Mumford, T. Shiota, Filtering, Segmentation and Depth. VIII, 143 pages. 1993.

Vol. 663: G. v. Bochmann, D. K. Probst (Eds.), Computer Aided Verification. Proceedings, 1992. IX, 422 pages. 1993.

Vol. 664: M. Bezem, J. F. Groote (Eds.), Typed Lambda Calculi and Applications. Proceedings, 1993. VIII, 433 pages. 1993.

Vol. 665: P. Enjalbert, A. Finkel, K. W. Wagner (Eds.), STACS 93. Proceedings, 1993. XIV, 724 pages. 1993.

Vol. 666: J. W. de Bakker, W.-P. de Roever, G. Rozenberg (Eds.), Semantics: Foundations and Applications. Proceedings, 1992. VIII, 659 pages. 1993.

Vol. 667: P. B. Brazdil (Ed.), Machine Learning: ECML – 93. Proceedings, 1993. XII, 471 pages. 1993. (Subseries LNAI).

Vol. 668: M.-C. Gaudel, J.-P. Jouannaud (Eds.), TAPSOFT '93: Theory and Practice of Software Development. Proceedings, 1993. XII, 762 pages. 1993.

Vol. 669: R. S. Bird, C. C. Morgan, J. C. P. Woodcock (Eds.), Mathematics of Program Construction. Proceedings, 1992. VIII, 378 pages. 1993.

Vol. 670: J. C. P. Woodcock, P. G. Larsen (Eds.), FME '93: Industrial-Strength Formal Methods. Proceedings, 1993. XI, 689 pages. 1993.

Vol. 671: H. J. Ohlbach (Ed.), GWAI-92: Advances in Artificial Intelligence. Proceedings, 1992. XI, 397 pages. 1993. (Subseries LNAI).

Vol. 672: A. Barak, S. Guday, R. G. Wheeler, The MOSIX Distributed Operating System. X, 221 pages. 1993.

Vol. 673: G. Cohen, T. Mora, O. Moreno (Eds.), Applied Algebra, Algebraic Algorithms and Error-Correcting Codes. Proceedings, 1993. X, 355 pages 1993.

Vol. 674: G. Rozenberg (Ed.), Advances in Petri Nets 1993. VII, 457 pages. 1993.

Vol. 675: A. Mulkers, Live Data Structures in Logic Programs. VIII, 220 pages. 1993.

Vol. 676: Th. H. Reiss, Recognizing Planar Objects Using Invariant Image Features. X, 180 pages. 1993.

Vol. 677: H. Abdulrab, J.-P. Pécuchet (Eds.), Word Equations and Related Topics. Proceedings, 1991. VII, 214 pages. 1993.

Vol. 678: F. Meyer auf der Heide, B. Monien, A. L. Rosenberg (Eds.), Parallel Architectures and Their Efficient Use. Proceedings, 1992. XII, 227 pages. 1993.

Vol. 679: C. Fermüller, A. Leitsch, T. Tammet, N. Zamov, Resolution Methods for the Decision Problem. VIII, 205 pages. 1993. (Subseries LNAI).

Vol. 680: B. Hoffmann, B. Krieg-Brückner (Eds.), Program Development by Specification and Transformation. XV, 623 pages. 1993.

Vol. 681: H. Wansing, The Logic of Information Structures. IX, 163 pages. 1993. (Subseries LNAI).

Vol. 682: B. Bouchon-Meunier, L. Valverde, R. R. Yager (Eds.), IPMU '92 – Advanced Methods in Artificial Intelligence. Proceedings, 1992. IX, 367 pages. 1993.

Vol. 683: G.J. Milne, L. Pierre (Eds.), Correct Hardware Design and Verification Methods. Proceedings, 1993. VIII, 270 Pages. 1993.

Vol. 684: A. Apostolico, M. Crochemore, Z. Galil, U. Manber (Eds.), Combinatorial Pattern Matching. Proceedings, 1993. VIII, 265 pages. 1993.

Vol. 685: C. Rolland, F. Bodart, C. Cauvet (Eds.), Advanced Information Systems Engineering. Proceedings, 1993. XI, 650 pages. 1993.

Vol. 686: J. Mira, J. Cabestany, A. Prieto (Eds.), New Trends in Neural Computation. Proceedings, 1993. XVII, 746 pages. 1993.

Vol. 687: H. H. Barrett, A. F. Gmitro (Eds.), Information Processing in Medical Imaging. Proceedings, 1993. XVI, 567 pages. 1993.

Vol. 688: M. Gauthier (Ed.), Ada - Europe '93. Proceedings, 1993. VIII, 353 pages. 1993.

Vol. 689: J. Komorowski, Z. W. Ras (Eds.), Methodologies for Intelligent Systems. Proceedings, 1993. XI, 653 pages. 1993. (Subseries LNAI).

Vol. 690: C. Kirchner (Ed.), Rewriting Techniques and Applications. Proceedings, 1993. XI, 488 pages. 1993.

Vol. 691: M. Ajmone Marsan (Ed.), Application and Theory of Petri Nets 1993. Proceedings, 1993. IX, 591 pages. 1993.

Vol. 692: D. Abel, B.C. Ooi (Eds.), Advances in Spatial Databases. Proceedings, 1993. XIII, 529 pages. 1993.

Vol. 693: P. E. Lauer (Ed.), Functional Programming, Concurrency, Simulation and Automated Reasoning. Proceedings, 1991/1992. XI, 398 pages. 1993.

Vol. 694: A. Bode, M. Reeve, G. Wolf (Eds.), PARLE '93. Parallel Architectures and Languages Europe. Proceedings, 1993. XVII, 770 pages. 1993.

Vol. 695: E. P. Klement, W. Slany (Eds.), Fuzzy Logic in Artificial Intelligence. Proceedings, 1993. VIII, 192 pages. 1993. (Subseries LNAI).

Vol. 696: M. Worboys, A. F. Grundy (Eds.), Advances in Databases. Proceedings, 1993. X, 276 pages. 1993.

Vol. 697: C. Courcoubetis (Ed.), Computer Aided Verification. Proceedings, 1993. IX, 504 pages. 1993.

Vol. 698: A. Voronkov (Ed.), Logic Programming and Automated Reasoning. Proceedings, 1993. XIII, 386 pages. 1993. (Subseries LNAI).

Vol. 699: G. W. Mineau, B. Moulin, J. F. Sowa (Eds.), Conceptual Graphs for Knowledge Representation. Proceedings, 1993. IX, 451 pages. 1993. (Subseries LNAI).

Vol. 700: A. Lingas, R. Karlsson, S. Carlsson (Eds.), Automata, Languages and Programming. Proceedings, 1993. XII, 697 pages. 1993.

Vol. 701: P. Atzeni (Ed.), LOGIDATA+: Deductive Databases with Complex Objects. VIII, 273 pages. 1993.

Vol. 702: E. Börger, G. Jäger, H. Kleine Büning, S. Martini, M. M. Richter (Eds.), Computer Science Logic. Proceedings, 1992. VIII, 439 pages. 1993.

Vol. 703: M. de Berg, Ray Shooting, Depth Orders and Hidden Surface Removal. X, 201 pages. 1993.